Study Guide for

Essentials of Anatomy & Physiology

First Edition

Drew Case, RN, MSN, APRN

3251 Riverport Lane
St. Louis, Missouri 63043

Study Guide for Essentials of Anatomy & Physiology
First Edition

ISBN: 978-0-323-07451-3

Notices

Knowledge and best practice in this field are constantly changing. As new research and experience broaden our understanding, changes in research methods, professional practices, or medical treatment may become necessary.

Practitioners and researchers must always rely on their own experience and knowledge in evaluating and using any information, methods, compounds, or experiments described herein. In using such information or methods they should be mindful of their own safety and the safety of others, including parties for whom they have a professional responsibility.

With respect to any drug or pharmaceutical products identified, readers are advised to check the most current information provided (i) on procedures featured or (ii) by the manufacturer of each product to be administered, to verify the recommended dose or formula, the method and duration of administration, and contraindications. It is the responsibility of practitioners, relying on their own experience and knowledge of their patients, to make diagnoses, to determine dosages and the best treatment for each individual patient, and to take all appropriate safety precautions.

To the fullest extent of the law, neither the Publisher nor the authors, contributors, or editors, assume any liability for any injury and/or damage to persons or property as a matter of products liability, negligence or otherwise, or from any use or operation of any methods, products, instructions, or ideas contained in the material herein.

The Publisher

Some material was previously published.

ISBN: 978-0-323-07451-3

Acquisitions Editor: Jeff Downing
Developmental Editor: Karen C. Turner
Editorial Assistant: Lauren Mussig
Publishing Services Manager: Jeffrey Patterson
Project Manager: Mary Stueck
Design Manager: Teresa McBryan

Printed in the United States of America
Last digit is the print number: 9 8 7 6 5 4 3 2 1

Acknowledgments

I dedicate this book to my wife Sheena, my son Wyatt, and my new baby girl Adree. I hope that my work someday inspires my children because they inspire me every day. I would also like to thank Steve Bassett for being a positive role model in the field of teaching anatomy and physiology and encouraging me to get involved with reviewing textbooks and publishing my own work. Lastly, I would like to thank Jeff Downing for giving me the opportunity to work on this project (hopefully this is the first of many) and also Karen Turner for all her help and patience walking me through the process. I hope the students find this helpful.

Drew Case

Introduction

This Study Guide was designed to help you learn, review, and most importantly, prepare for exams for Essentials of Anatomy & Physiology. Prior to using this Study Guide, it is my strong recommendation that you have had the content in lecture, read the corresponding chapters in the text, and that you have studied the material. The best way to check your knowledge and preparedness prior to taking an exam is by taking a practice exam. I have designed this Study Guide so that you can quickly, efficiently, and effectively check your level of comprehension of the material prior to exams. The Study Guide is set up in such a manner that you will be able to quickly go back to the textbook and study or review more should you have difficulty answering the questions. I based the format of this guide on what would be most likely encountered on course exams. The exam is the final test of our efforts and success of learning the material.

I will give you study tips, hints, and suggestions throughout the Study Guide. They are based on years of teaching and student feedback and are worth reading and giving a try. You may not find all of them helpful. Learning is a "trial-and-error" process. Be open to new ways of studying and try different methods and you will find the way that works best for you.

I have organized the questions of each chapter in the same pattern to simplify studying and create predictability in the Study Guide. Below is a list and description of the questions in the order that will be encountered throughout the guide. Not all of the question types will always be used for each section.

Fill in the Blanks

Fill in the blank questions are the highest level of difficulty and require you to recall the answer from memory. They are very similar many times to multiple-choice questions simply without the choices, thus the higher level of difficulty. They make excellent questions to put on flash cards for studying.

Matching

Matching questions are one of the lowest levels of difficulty of the questions provided and may contain some of the more difficult definitions to increase the challenge. Match the term on the left with the definition on the right and write in the letter of the term in the answer blank next to the term. Each term can only be used once unless noted otherwise. I have tried to make the answers short and to the point. Sometimes I use words or phrases that I found helpful in learning the material.

Multiple Choice

Multiple-choice questions are a moderate difficulty level and are the most frequently encountered type of question on exams and more importantly, board and certification exams. Multiple-choice questions are short answer or fill in the blank questions with choices. If you have difficulty with multiple-choice questions because you "talk yourself out of the right answer," cover the choices and answer it like a fill in the blank and then see if your answer is one of the choices. If so, pick that answer and move on.

Labeling

Labeling questions are a completely different way of testing your knowledge from all the other question types presented. This is referred to as "visual learning" and is an excellent method of testing what is "memorized." I recommend using a scrap piece of paper to record your answers so that you may label the pictures/diagrams over and over till you can consistently get a 100%.

Short Answer

Short answers are similar to fill in the blank in difficulty and style. These questions usually require listing out multiple answers and often in order. These too also make excellent flash card questions. The majority of the short answer questions cover critical material that is very likely to show up on exams.

Study Tips

Every chapter will have study tips that include things like mnemonics and other ideas on how to memorize and learn the material. The study tips will have an icon next to them to help identify them and make them stand out. These study tips are based on years of teaching and positive student feedback, so be sure to read them.

Sample Flash Cards

FC This symbol, used throughout the Study Guide, indicates questions or information that would be helpful to include in making flash cards. It does not suggest making a flash card with the same question or format but rather that the term, concept, structure(s), or definition has been frequently included on exams covering this material.

Flash cards can be one of the most effective methods of putting vast amounts of material to memory. The great advantage of using flash cards is that you are testing yourself just as though you had your own personal tutor sitting there asking you questions. Think about what you would want someone to ask you to help you prepare for an exam, and that is what you put on the flash card. If you make flash cards correctly, you remove the doubt or question, "do I really know the material?" You either answer the question right or wrong. Here are just a few quick tips for making flash cards:

1. Keep them simple; use as few words as possible
2. Question on one side (be sure to ask a question), answer on the other
3. DO NOT make flash cards on material that you already know or have memorized
4. Make a flash card on everything you think will be on the exam that you DO NOT have memorized
5. Limit the number of questions to only a couple per card based on the complexity of the answer. The questions and answers should be short and to the point like fill in the blank or multiple-choice questions without the choices. AVOID long answer questions that take more than one sentence to answer.
6. List questions are excellent flash card questions as are matching questions.

Here are a couple of sample flash cards from Chapter 1:

1. Study of human body's structures?
2. Standing with feet and head facing forward and the arms at the sides with the palms facing forward?
3. Study of body's functions?
4. Study of body's dysfunction?

(Notice how short and simple the questions are? Answer all the questions on this side in your head or out loud and turn the card over.)

1. Anatomy
2. Anatomical position
3. Physiology
4. Pathophysiology

(Answers are short; find these questions in Chapter 1. Several were multiple-choice questions without the choices. I included four questions on this one card because the answers were so short.)

Here is how you can make questions from the matching sections:

(side one)

Define:

1. Dorsal
2. Ventral
3. Otic
4. Distal
5. Medial

(side two)

1. Toward the back or posterior
2. Toward the front or anterior
3. Ear
4. Farther away from attachment to trunk
5. Toward the midline of the body

Hopefully this helps you get started. Making flash cards is an art and takes practice. If your cards are not helping you, then you most likely are doing them wrong. If they are taking too long to make, you are making them wrong. It should take you less than an hour to make flash cards for one chapter. DO NOT make flash cards until you have had the subject in lecture first. You don't want to make cards for material that was not covered or is not going to be on the exam. Good luck!

Contents

CHAPTER 1

Organization of the Human Body

HOW TO APPROACH THE ORGANIZATION OF THE HUMAN BODY

I cannot emphasize enough how important this chapter is! The information you learn in this chapter you will use throughout the entire book. Many exams are multiple-choice with few to no pictures provided. How can we test you on bone structures without pictures? By using the anatomical terms and directions covered in this chapter. If you do not fully grasp all of the anatomical terms and directions before moving on, it will catch up with you later, so be sure you are **100%** clear on all of them before moving on. Take a quick look at some of the multiple-choice questions under bones and muscles to see how they incorporate this chapter. This will give you an idea of how well you will need to know them. The body cavities and abdominopelvic regions and quadrants are also frequently used throughout the text and more importantly in life. You will hear many of these terms used when you or your family are being seen by medical professionals. It is very helpful to know what they are talking about. Pay particular attention to *medial/lateral* and *proximal/distal*. These terms are used consistently and frequently give students difficulty.

FC This symbol indicates questions or information that would be helpful to include in making flash cards. It does not suggest making a flash card with the same question or format but rather that the term, concept, structure(s), or definition has been frequently included on exams covering this material.

STUDY TIP

If you use a piece of scratch paper to write your answers on, you can test yourself over and over again. Try highlighting the numbers of the ones you missed so you know which ones to test on again. Keep repeating this procedure until you are able to answer them all. Now you're ready for a test. DO THIS FOR EVERY CHAPTER! The BEST way to prepare for a test is taking tests!

CHARACTERISTICS OF LIFE AND LEVELS OF ORGANIZATION

Fill in the blanks with the correct answer.

1. Atoms bond together making ____________________, which combine in specific ways making larger and more complex molecules called ____________________.

2. ____________________ are "tiny organs" that keep cells alive.

3. ____________________ are the foundation and basic units of our bodies.

4. Cells are organized into ____________________, which make up ____________________, which make up ____________________, which finally make the human ____________________.

Matching – match each definition to the correct term.

	Term		Definition
5. ______	Responsiveness	a.	Organized increase in size
6. ______	Conductivity	b.	Ability to sense, monitor, and respond to changes in environment
7. ______	Digestion	c.	Exchange of gases between organism and environment
8. ______	Growth	d.	Movement of molecules through a membrane
9. ______	Reproduction	e.	Capacity to transmit a wave of excitation from one point to another
10. ______	Absorption	f.	Production and release of important substances
11. ______	Respiration	g.	Removal of waste
12. ______	Secretion	h.	Breaking down of complex to simpler substances
13. ______	Circulation	i.	Movement of body fluids through hollow vessels
14. ______	Excretion	j.	Formation of new individuals

Multiple choice – select the best answer.

FC 15. The study of the human body's structures is called:
- a. physiology.
- b. anatomy.
- c. pathophysiology.
- d. chemistry.

FC 16. If a body is standing with feet and head facing forward and the arms at the sides with the palms facing forward, it is said to be what?
- a. Anatomically correct
- b. In anatomical position
- c. In anatomy stance
- d. In dissection posture

17. Which of the following are the smallest structural units that exhibit the basic characteristics of life?
- a. Tissues
- b. Atoms
- c. Organelles
- d. Cells

18. Which of the following is the correct sequence of levels of organization?
    a. Cells, organelles, tissues, organs, organ systems
    b. Elements, atoms, molecules, macromolecules, organelles
    c. Macromolecules, molecules, tissues, organelles, cells
    d. Atoms, elements, molecules, macromolecules, organelles

FC 19. *Physiology* refers to the:
    a. ability to produce chemical and electrical reactions.
    b. function of the human body.
    c. study of disease in the human body.
    d. study of the structures of the body.

BODY CAVITIES AND REGIONS

Fill in the blanks with the correct answer.

20. The ____________________ cavity includes the ____________________ and spinal cavities.

21. The ventral cavity includes the ____________________ cavity and the ____________________ cavity.

22. The ____________________ cavity includes the right and left ____________________ cavity and the central portion called the ____________________, which contains the heart, trachea, esophagus, and many other structures.

23. When discussing membranes and body cavities, ____________________ refers to covering the inside wall of a body cavity, whereas ____________________ refers to covering the organ inside the cavity.

24. A thin layer of ____________________ is located between membranes in body cavities, lubricating the organs.

Labeling – label the following diagrams.

25. Major body cavities

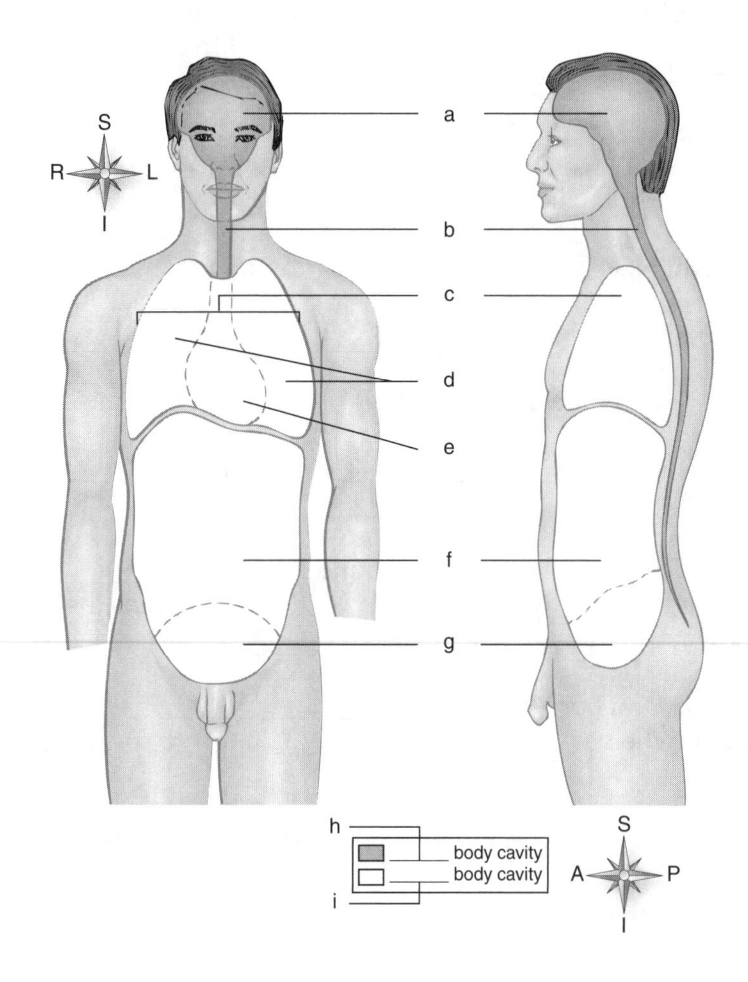

a. ______________________________

b. ______________________________

c. ______________________________

d. ______________________________

e. ______________________________

f. ______________________________

g. ______________________________

h. ______________________________

i. ______________________________

26. Nine regions of the abdominopelvic cavity

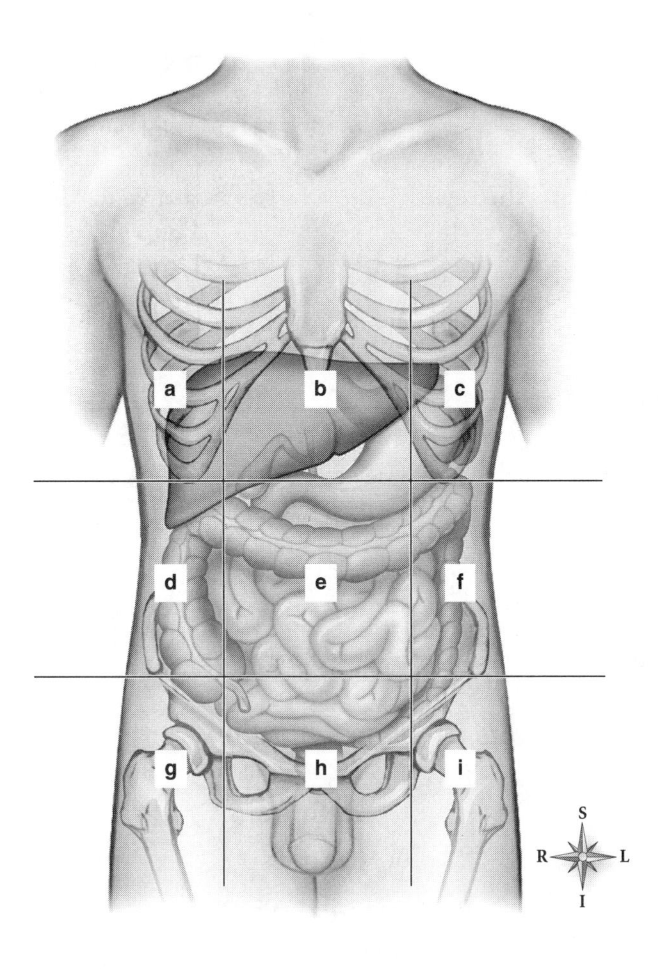

a. ______________________

b. ______________________

c. ______________________

d. ______________________

e. ______________________

f. ______________________

g. ______________________

h. ______________________

i. ______________________

Multiple choice – select the best answer.

27. What cavity is directly medial to the right pleural cavity?
    a. Pelvic
    b. Cranial
    c. Ventral
    d. Mediastinum

28. The body region that contains the head, neck, and trunk is the:
    a. appendicular.
    b. ventral.
    c. dorsal.
    d. axial.

FC 29. *Visceral* refers to:
    a. internal environment.
    b. inside wall of body cavities.
    c. internal organs.
    d. lubrication.

30. The cranial and spinal cavities comprise the:
    a. ventral cavity.
    b. mediastinum.
    c. dorsal cavity.
    d. axial cavity.

31. The lungs are found in which cavity?
    a. Pleural
    b. Dorsal
    c. Mediastinum
    d. Parietal

ANATOMICAL TERMS, BODY PLANES AND SECTIONS

FC *Matching – match each description with the correct term.*

	Term		Definition
32. ______	Popliteal	a.	Forearm
33. ______	Mammary	b.	Upper part of arm
34. ______	Antebrachial	c.	Wrist
35. ______	Brachial	d.	Hip
36. ______	Carpal	e.	Elbow
37. ______	Coxal	f.	Chin
38. ______	Cubital	g.	Ear
39. ______	Mental	h.	Foot
40. ______	Otic	i.	Breast
41. ______	Pedal	j.	Behind knee

FC *Matching – match each description with the correct term.*

	Term		Definition
42. ______	Dorsal	a.	Farther away from attachment to trunk
43. ______	Medial	b.	Nearer the surface
44. ______	Apical	c.	Inner region of an organ
45. ______	Superficial	d.	Means lower or below
46. ______	Cortical	e.	Base or widest part of an organ
47. ______	Medullary	f.	Away from the midline of the body
48. ______	Basal	g.	Narrower tip of an organ
49. ______	Lateral	h.	Also called posterior
50. ______	Distal	i.	Outer region or layer of an organ
51. ______	Inferior	j.	Toward the midline of the body

STUDY TIP

Multiple-choice questions are excellent examples of what you might want to put on flash cards or how to format material on the cards.

When you find multiple-choice questions that are very important or contain information you need to know, you can simply write the question on one side of the card and the answer on the other. A multiple-choice question is simply a fill in the blank or short answer with choices. It is also a good idea to share your flash cards with other students in the class to share ideas. The BEST flash cards are the ones you make. It takes some practice, but it's worth your time.

Multiple choice – select the best answer.

52. What abdominopelvic region is lateral and superior to hypogastric?
    a. Epigastric
    b. Right lumbar
    c. Umbilical
    d. Left iliac

53. Which of the following refers to fingers and toes?
    a. Digital
    b. Carpal
    c. Tarsal
    d. Hallux

54. What plane of dissection divides the body into dorsal and ventral halves?
    a. Coronal
    b. Sagittal
    c. Midsagittal
    d. Transverse

55. Orbital is what to occipital?
    a. Inferior and posterior
    b. Anterior and proximal
    c. Ventral and superior
    d. Inferior and ventral

56. *Mental* refers to the what?
    a. Head
    b. Side of skull
    c. Neck
    d. Chin

57. Hallux is what to popliteal?
    a. Proximal and inferior
    b. Medial and distal
    c. Lateral and distal
    d. Proximal and medial

58. If the cephalic was separated from the cervical, what plane of dissection was used?
    a. Coronal
    b. Frontal
    c. Transverse
    d. Sagittal

59. The crural is __________ to the sural.
    a. dorsal
    b. ventral
    c. proximal
    d. distal

60. Which of the following is anterior, inferior, and medial to the otic?
    a. Occipital
    b. Frontal
    c. Mental
    d. Orbital

61. The olecranal is dorsal to the what?
    a. Sural
    b. Gluteal
    c. Patellar
    d. Occipital

Labeling – label the following diagrams.

62. Specific body regions, ventral view

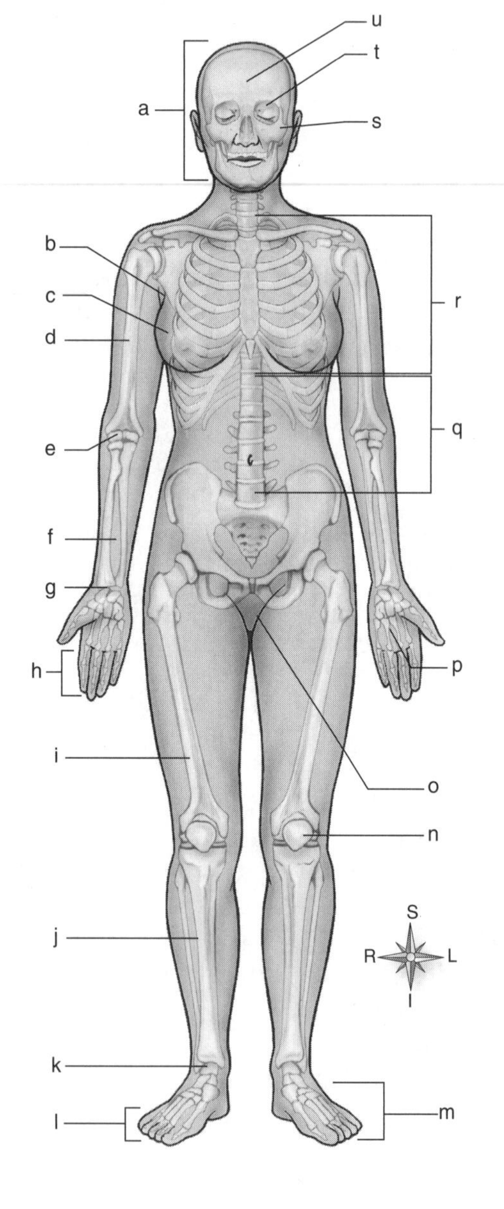

a. ______________________________
b. ______________________________
c. ______________________________
d. ______________________________
e. ______________________________
f. ______________________________
g. ______________________________
h. ______________________________
i. ______________________________
j. ______________________________
k. ______________________________
l. ______________________________
m. ______________________________
n. ______________________________
o. ______________________________
p. ______________________________
q. ______________________________
r. ______________________________
s. ______________________________
t. ______________________________
u. ______________________________

63. Specific body regions, dorsal view

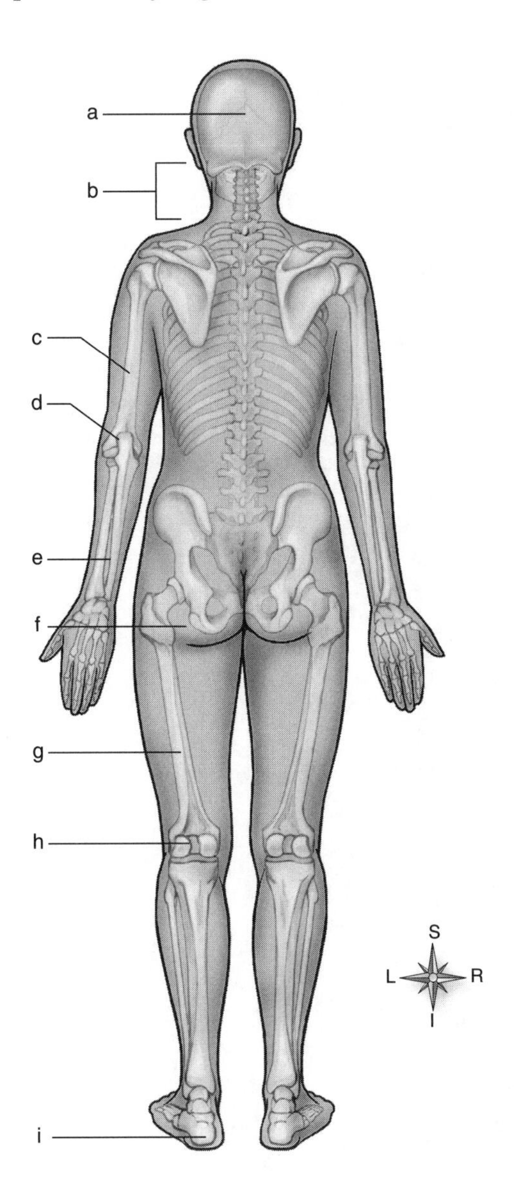

a. ______________________________

b. ______________________________

c. ______________________________

d. ______________________________

e. ______________________________

f. ______________________________

g. ______________________________

h. ______________________________

i. ______________________________

64. Directions and planes of the body

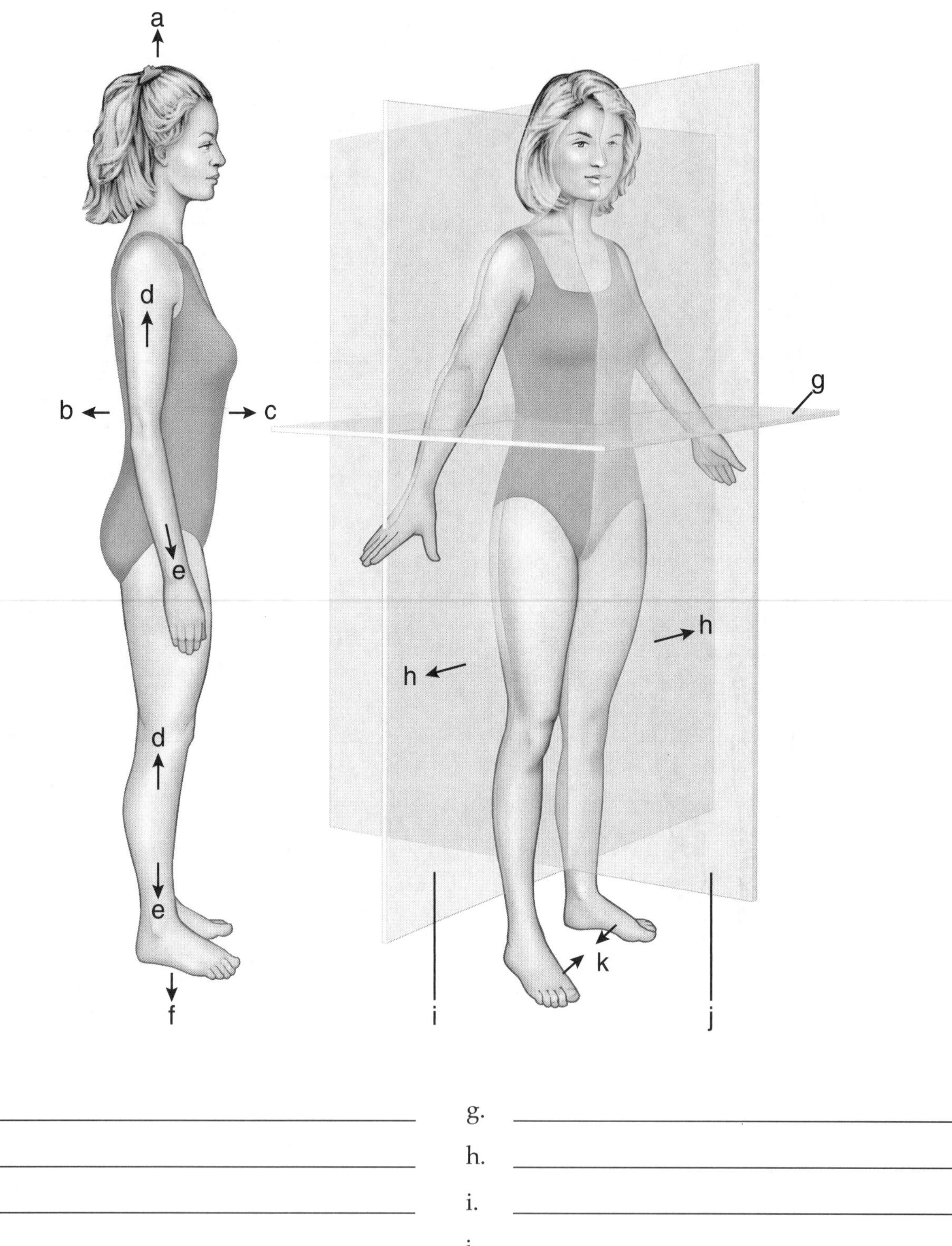

a. ______________________

b. ______________________

c. ______________________

d. ______________________

e. ______________________

f. ______________________

g. ______________________

h. ______________________

i. ______________________

j. ______________________

k. ______________________

HOMEOSTASIS

Fill in the blanks with the correct answer.

65. ____________________ describes a state of internal consistency or "staying the same."

66. The processes that maintain or restore homeostasis are called ____________________ and they accomplish this self-regulation through a control system called a(n) ____________________.

67. The four basic components in every feedback loop are ____________________, ____________________, ____________________, and ____________________.

68. When discussing feedback loops, ____________________ means the signal traveling toward, whereas ____________________ means the signal moving away from a particular center.

69. ____________________ feedback control systems are always inhibitory, whereas ____________________ feedback is always stimulatory.

70. Negative feedback control produces an action that is ____________________ to the change that activated the system and ____________________ physiological variables.

71. A positive feedback control system produces an action that ____________________ the change that activated the system.

CHAPTER 2

The Chemistry of Life

HOW TO APPROACH THE CHEMISTRY OF LIFE

I have found the most useful and important part of chemistry is understanding the atom and how it relates to ions. The atom is similar to a small solar system, with the nucleus (containing the protons and neutrons) being the sun and the electrons being the planets in orbit around the sun. Pay particular attention to the electron shells or orbits. The protons and electrons have charges. These charges determine the overall charge of the atom. The overall charge of the atom will determine how all atoms will react with each other (chemical bonds). Understanding the atom and its charge will also explain the importance of electrolytes and ions. If you have a solid understanding of the atom, subatomic particles, charges, electron shells or orbits, and chemical bonds, the rest of the chapter will fall into place and you should have little difficulty mastering the other concepts.

BASIC CHEMISTRY

Fill in the blanks with the correct answer.

1. A ____________________ reaction takes place when two or more atoms or molecules interact to form a new chemical bond and a new compound.

FC 2. A ____________________ reaction takes place when two or more atoms bond to create a molecule.

3. In ____________________ reactions, bonds within a reactant molecule break and produce simpler molecules usually releasing energy.

4. ____________________ reactions break down two compounds and synthesize two new compounds.

5. ____________________ refers to all chemical reactions that occur in the body.

FC ***Matching – match each description with the correct term.***

	Term		**Definition**
6. ______	Ionic bond	a.	Same element with different number of neutrons
7. ______	Covalent bond	b.	Sum of the protons and average number of neutrons in the nucleus
8. ______	Hydrogen bond	c.	Number of protons in nucleus
9. ______	Isotope	d.	Subatomic particle
10. ______	Atomic weight	e.	Two or more atoms joined chemically
11. ______	Atomic number	f.	Sharing electrons
12. ______	Proton	g.	"Pure" substances
13. ______	Compound	h.	Transferring electrons
14. ______	Elements	i.	Anything that takes up space and has mass
15. ______	Matter	j.	Not true bonds

Multiple choice – select the best answer.

16. Which elements comprise about 96% of the human body?
    a. Oxygen, hydrogen, sodium, and potassium
    b. Carbon, hydrogen, oxygen, and nitrogen
    c. Calcium, carbon, iron, and oxygen
    d. Sodium, calcium, oxygen, and carbon

17. Which of the following is an example of an element?
    a. Water
    b. Steel
    c. Gold
    d. Air

18. The principle in chemical interactions that causes atoms to gain, lose, or share electrons is called the:
    a. bond rule.
    b. energy rule.
    c. octet rule.
    d. shell rule.

19. When an atom loses or gains electrons from another atom, or when two or more atoms share electrons, it forms a:
    a. chemical bond.
    b. ionic bond.
    c. covalent bond.
    d. hydrogen bond.

20. This subatomic particle is frequently considered to have no mass.
    a. Proton
    b. Neutron
    c. Electron
    d. Centron

21. Electrons are arranged in ________ around the nucleus.
    a. orbital levels
    b. shells
    c. octets
    d. nuclear orbits

22. Electrons shared between atoms that result in an uneven charge around the molecule are called:
    a. unstable.
    b. catabolism.
    c. polar.
    d. ionic.

23. How many electrons does it take to fill the second and third shells of an atom?
    a. 2
    b. 4
    c. 6
    d. 8

24. Which type of bond is formed due to polarity?
   a. Chemical
   b. Ionic
   c. Covalent
   d. Hydrogen

25. Which type of bond results from sharing?
   a. Molecular
   b. Ionic
   c. Covalent
   d. Hydrogen

26. If the atoms in a molecule are from different elements, the molecule is called a:
   a. compound.
   b. isotope.
   c. mixture.
   d. ion.

27. Which subatomic particle has mass and can vary from atom to atom?
   a. Electron
   b. Neutron
   c. Proton
   d. Ion

28. Energy is usually released in these types of reactions and simpler molecules are produced.
   a. Decomposition
   b. Exchange
   c. Catabolic
   d. Synthesis

29. Catabolism reactions generally result in the production of:
   a. ions.
   b. ATP.
   c. complex molecules.
   d. proteins.

30. A synthesis reaction takes place when:
   a. bonds within a reactant molecule break and produce simpler molecules.
   b. two compounds are broken down and synthesize two new compounds.
   c. neutrons are taken from one molecule and added to another.
   d. when two or more atoms interact to form new chemical bonds and compounds.

Short answer

31. What are the three basic types of chemical reactions?

FC 32. Metabolism is divided into what two groups?

FC 33. List the three states of matter.

FC 34. List the subatomic particles.

FC 35. List the three types of chemical bonds discussed thus far.

INORGANIC MOLECULES

Multiple choice – select the best answer.

FC 36. What is something that can dissolve various substances?
a. Solute
b. Electrolyte
c. Solution
d. Base

37. What accepts hydrogen ions and are thus referred to as *proton acceptors*?
a. Acids
b. Bases
c. Solutions
d. Inorganic molecules

38. The pH scale measures the relative concentration of what?
a. Hydrogen ions
b. Oxygen ions
c. Potassium ions
d. Water

39. A pH of 1 is how many times more acidic than a pH of 4?
a. 4
b. 400
c. 10,000
d. 1000

FC 40. A substance that minimizes changes in concentrations of H^+ and OH^- in a solution is called a:
a. salt.
b. cation.
c. anion.
d. buffer.

41. All of the following are inorganic molecules EXCEPT:
a. oxygen.
b. electrolytes.
c. salts.
d. lipids.

42. If you add normal table salt to a glass of pure water, the salt will become the:
a. solution.
b. solute.
c. solvent.
d. organic molecule.

FC 43. What is called the "universal solvent"?
a. Acid
b. Alcohol
c. Water
d. Plasma

44. What compounds dissociate in a solution to form charged ions?
a. Electrolytes
b. Organic compounds
c. Inorganic compounds
d. Buffers

ORGANIC MOLECULES

Matching – match each term with the correct description.

	Term		Definition
45. ______	Monosaccharide	a.	Composed of a phosphate, sugar, and base
46. ______	Lipids	b.	Links amino acids together
47. ______	Peptide bond	c.	Fundamental energy-storage molecule
48. ______	Hydrolysis	d.	Breaks peptides down
49. ______	Disaccharide	e.	Makes polypeptides
50. ______	Nucleotides	f.	Protein accelerators or catalysts
FC 51. ______	ATP	g.	Hydrophobic
FC 52. ______	Enzyme	h.	Chain of monosaccharides
53. ______	Dehydration synthesis	i.	Sucrose
54. ______	Polysaccharide	j.	Glucose

STUDY TIP

The short answer questions are an excellent example of a good format for flash cards as well. Example – Put "List the four bases of DNA" on side A and the answer on side B. When the answer consists of four or more items per question, I recommend one question per card. The biggest mistake when making flash cards is putting too much information on one card. Remember they are called "FLASH" cards; besides, they are cheap, so make a bunch.

Short answer

FC 55. List two nucleic acids.

FC 56. What are the four bases of DNA?

57. What four elements make up protein?

58. Name the four major groups of organic compounds.

59. List the four major groups of lipids discussed.

CHAPTER 3

Anatomy of Cells

HOW TO APPROACH THE ANATOMY OF CELLS

A cell is similar to you. The cell is covered with a plasma membrane and you are covered with skin; both allow things to move in and out (semipermeable or selectively permeable). The nucleus is your head, containing vital information and you cannot live without it. The cytoplasm is your body, containing all of your organs/organelles vital to everyday normal function. Just like with the organs of the body, you must know all the organelles and their location and function. Also know the nucleus and ALL of its contents and their functions. The two matching sections below will have a lot of the basic information just discussed.

THE FUNCTIONAL ANATOMY OF CELLS

Multiple choice – select the best answer.

FC 1. The cytoplasm can be divided into what two parts?
   a. Water and solids
   b. Plasma membrane and nuclear membrane
   c. Cytosol and organelles
   d. Hydrophilic and hydrophobic

2. DNA is located where in the cell?
   a. Nucleus
   b. Nucleolus
   c. Cytosol
   d. Cytoplasm

3. Phospholipids and cholesterol, studded with a variety of proteins, glycoproteins, and glycolipids, comprise the:
   a. cell wall.
   b. nuclear membrane.
   c. plasma membrane.
   d. cytoplasmic membrane.

4. What is added to the phospholipid bilayer to strengthen it without losing flexibility?
   a. Cholesterol
   b. Protein
   c. Glycoproteins
   d. Glycolipids

5. Which of the following serve as identification markers on our cells identifying the cell as "us" or "me"?
   a. Cholesterol
   b. Rafts
   c. Glycolipids
   d. Glycoproteins

6. Which of the following is NOT a function of the cell membrane?
   a. Acts as receptor to trigger chemical reactions and other metabolic changes
   b. Identifies and destroys bacterial and cancerous cells
   c. Carries substances into the cytoplasm from outside the cell
   d. Prevents the body from accepting certain types of blood

BASIC ORGANELLES AND THEIR FUNCTIONS

Fill in the blanks with the correct answer.

7. Organelles are divided into the two broad categories of ____________________ and ____________________.

8. The ____________________ is a network of canals and sacs made of cell membrane. There are two types of this structure: the ____________________ and the ____________________.

9. Small membranous sacs called ____________________ contain enzymes that remove toxic chemicals from the cytosol. They are found in large numbers in the kidney and liver.

FC ***Matching – match each description with the correct term.***

	Term		Definition
10. ______	Rough endoplasmic reticulum	a.	"Digestive bags" or "cellular garbage disposals"
11. ______	Ribosomes	b.	Processes and packages proteins from ER
12. ______	Organelles	c.	"Power plants" of the cell
13. ______	Smooth endoplasmic reticulum	d.	Extends from the nucleus throughout the cytosol and has ribosomes attached all over it
14. ______	Golgi apparatus	e.	"Little organs"
15. ______	Lysosomes	f.	Break down abnormal and misfolded proteins
16. ______	Peroxisomes	g.	Site of protein synthesis, translate the genetic code into proteins
17. ______	Proteasomes	h.	Found mostly in liver and kidney
18. ______	Mitochondria	i.	Synthesizes membrane lipids, steroids, and some carbohydrates

Multiple choice – select the best answer.

19. Which of the following increase(s) the surface area of the inner membrane of mitochondria?
   a. Endoplasmic reticulum
   b. Cisternae
   c. Proteasomes
   d. Cristae

20. What are called "picky eaters" and destroy proteins one at a time?
   a. Proteases
   b. Proteasomes
   c. Lysosomes
   d. Enzymes

21. Which of the following organelles has its own DNA?
    a. Ribosome
    b. Golgi apparatus
    c. Mitochondrion
    d. Nucleolus

22. Steroid hormones originated where?
    a. Golgi apparatus
    b. Rough endoplasmic reticulum
    c. Smooth endoplasmic reticulum
    d. Ribosomes

23. "Free" or unattached ribosomes in the cytoplasm make proteins for:
    a. intracellular use.
    b. hormones.
    c. plasma membrane.
    d. transport to other cells.

FC 24. Small bubbles of endoplasmic reticulum pinch off, forming:
    a. vesicles.
    b. cisternae.
    c. lysosomes.
    d. plasmic vessels.

NUCLEUS, CYTOSKELETON, AND CELL CONNECTIONS

FC *Matching – match each description with the correct term.*

	Term		Definition
25. ______	Centrosomes	a.	Produce a waving-like motion
26. ______	Centrioles	b.	Coordinate the building and breaking of microtubules
27. ______	Microvilli	c.	Primarily made of RNA
28. ______	Cilia	d.	"Spot welds" that hold adjacent cells together
29. ______	Nucleus	e.	Membrane channels that join cytoplasm of two cells
30. ______	Chromatin	f.	"Control center" of the cell
31. ______	Nucleolus	g.	Long, unwound strands of DNA
32. ______	Chromosomes	h.	Tightly coiled and replicated DNA
33. ______	Desmosomes	i.	Important in formation of cellular extensions and role in cell division
34. ______	Gap junctions	j.	Serve to increase surface area

Multiple choice – select the best answer.

35. What holds sheets of cells together, making it difficult to permeate the cracks between cells?
    a. Desmosomes
    b. Gap junctions
    c. Tight junctions
    d. Sheet junctions

36. What forms "tunnels" that join the cytoplasm of two cells by fusing the two plasma membranes into a single coherent structure?
    a. Gap junctions
    b. Desmosomes
    c. Pores
    d. Tight junctions

37. Provides structural support for the cell and its components and a means of physically moving the cell.
    a. Organelles
    b. Plasma membrane
    c. Cytoskeleton
    d. Cytoplasm

FC 38. These molecules contain the "master code" for making all the RNA, enzymes, and proteins of a cell.
    a. DNA
    b. Nucleoplasm
    c. Nucleolus
    d. Ribosomes

39. Many protein subunits oriented in a spiral fashion are called:
    a. microfilaments.
    b. intermediate filaments.
    c. microtubules.
    d. glycoproteins.

40. Twisted strands of protein molecules often lying parallel are called:
    a. microfilaments.
    b. intermediate filaments.
    c. microtubules.
    d. glycoproteins.

41. Which of the following are important in the formation of cellular extensions such as microvilli, cilia, and flagella?
    a. Centrosomes
    b. Centrioles
    c. Microfilaments
    d. Intermediate filaments

42. Which of the following cell connections is most common in skin cells?
    a. Desmosome
    b. Gap junction
    c. Tight junction
    d. Sheet junction

43. Which of the following cell connections is most common in heart muscle?
    a. Desmosome
    b. Gap junction
    c. Tight junction
    d. Sheet junction

44. Which of the following cell connections is most common in the lining of the intestines?
    a. Desmosome
    b. Gap junction
    c. Tight junction
    d. Sheet junction

Labeling – label the following diagram.

45. Typical, or composite, cell

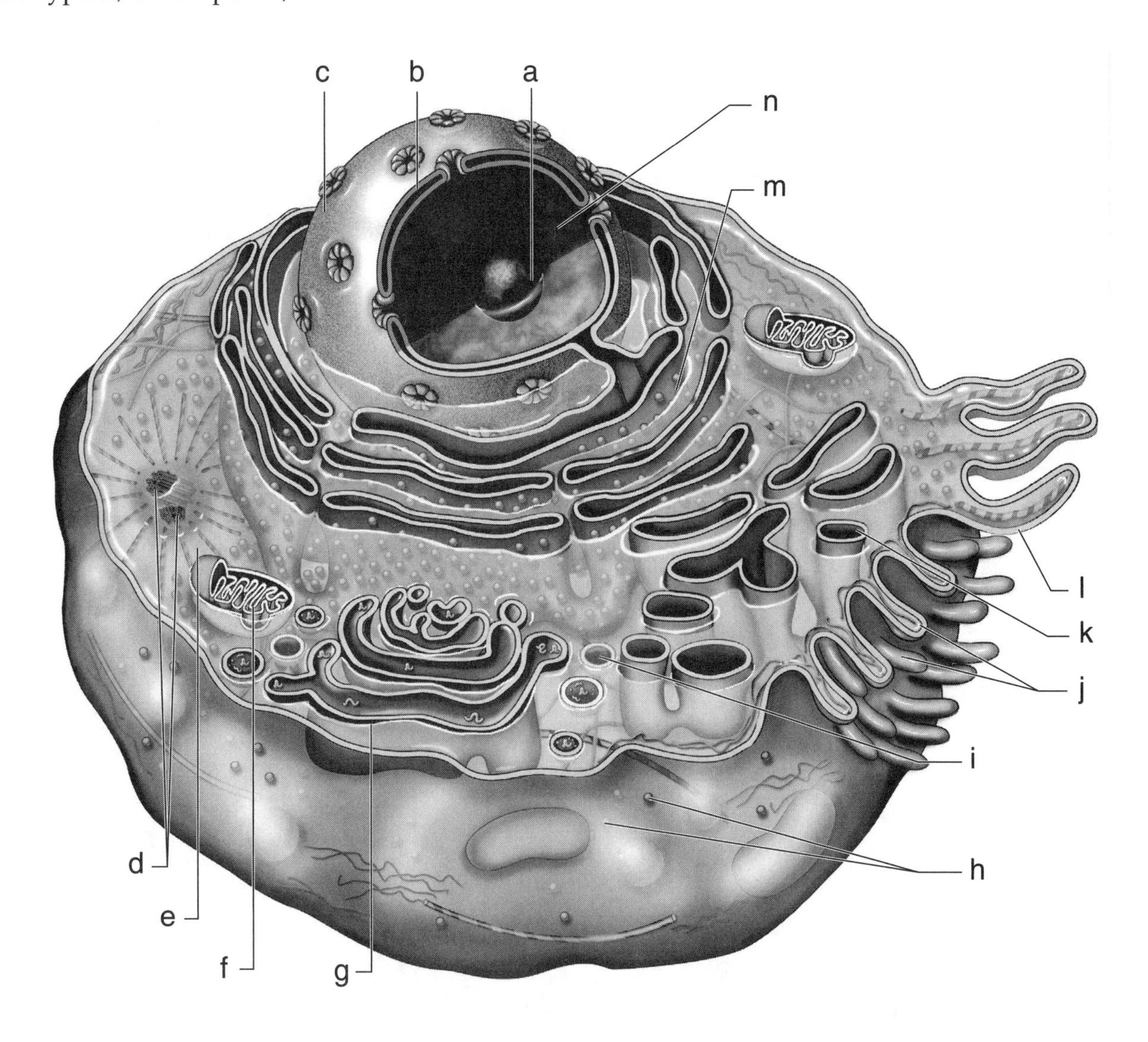

a. ______________________

b. ______________________

c. ______________________

d. ______________________

e. ______________________

f. ______________________

g. ______________________

h. ______________________

i. ______________________

j. ______________________

k. ______________________

l. ______________________

m. ______________________

n. ______________________

CHAPTER 4

Physiology of Cells

HOW TO APPROACH THE PHYSIOLOGY OF CELLS

Active and especially passive transport are very important concepts and you need to know what they are and all of the specific types of each. Diffusion is one of the most important concepts in this chapter because it is used so frequently in other chapters. Osmosis is a type of diffusion that is also mentioned repeatedly, so it is vital to understand it now rather than later. Students frequently have difficulty with osmosis, so be sure to see the study tip below. Remember, both diffusion and osmosis move down a concentration gradient or move from HIGH CONCENTRATION to LOW CONCENTRATION. Active transport is frequently used to move things against the concentration gradient. Diffusion and osmosis will explain things like breathing, IV solutions, dehydration/rehydration, laxatives and diarrhea, edema, nasal flushes, and much more. Do you see their importance now?

Cell metabolism and respiration are very helpful in understanding exercise, nutrition, and weight management. Most people do not understand the difference between aerobic and anaerobic. Make sure you know how they are similar and how they are different. Where they take place and what their products and byproducts are. Why do you need to drink a lot of fluid after anaerobic exercise to keep from getting sore? See the study tip regarding aerobics. When you are done with this chapter, try to think of how this information can be applied to your exercise program, nutrition, and rehabilitation.

Examples: When you are jogging, are you aerobic or anaerobic? What is best to drink after a long, hard workout and why (osmosis)? To decrease your risks for heart disease, do you need more aerobic or anaerobic exercise? Why does drinking soda make you thirstier? Why do many bars give away free peanuts and popcorn (osmosis)? Who said learning can't be fun and useful?

MOVEMENT OF SUBSTANCES THROUGH CELL MEMBRANES

Fill in the blanks with the correct answer.

1. ____________________ require energy to transport molecules against their concentration gradient.

2. The sodium-potassium pump exchanges (number of ions) ___________ sodium ions for ___________ potassium ions using the energy-containing molecule called ____________________.

3. ____________________ and ____________________ are always *passive processes* in which the energy for transport across a membrane comes from molecular movements and collisions, and not from the cell.

FC 4. ____________________ refers to lower potential osmotic pressure, whereas ____________________ refers to higher osmotic pressure.

FC ***Matching – match each description with the correct term.***

	Term		Definition
5. ______	Endocytosis	a.	Diffusion and osmosis
6. ______	Active transport	b.	Moving from high to low concentration
7. ______	Diffusion	c.	More water and less solutes
8. ______	Osmosis	d.	Less water and more solutes
9. ______	Pinocytosis	e.	Phagocytosis and pinocytosis
10. ______	Hypotonic	f.	Endocytosis and exocytosis
11. ______	Hypertonic	g.	Means "cell eating"
12. ______	Passive transport	h.	Means "cell drinking"
13. ______	Exocytosis	i.	Involves Golgi apparatus packaging material in vesicles
14. ______	Phagocytosis	j.	Water moving from high to low concentration across a membrane

STUDY TIP

FC When trying to understand osmosis, it is easier to focus on where the highest concentration of solutes is. Focus on the solutes rather than the water concentration. The fluid with the highest concentration of solutes is the hypertonic solution. This means the other must be the hypotonic. Thus you only have to figure out one to get the other. The other way to remember the direction of water movement is "Hyper – PULLS." This means "Hypo – PUSHES." Water always moves into or toward the hypertonic solution due to the solute concentration.

Multiple choice – select the best answer.

15. Which of the following correctly describes a hypertonic extracellular fluid (ECF)?
    a. The cell contains a higher concentration of solutes than the intracellular fluid (ICF).
    b. The ICF contains a higher concentration of solutes than the ECF.
    c. The ICF consists of a lower concentration of water than the ECF.
    d. The ECF contains a higher concentration of solutes than ICF.

16. Which of the following correctly describes a hypotonic ICF?
    a. The cell contains a higher concentration of solutes than the ICF.
    b. The ICF contains a higher concentration of solutes than the ECF.
    c. The ICF consists of a lower concentration of water than the ECF.
    d. The ECF contains a higher concentration of solutes than ICF.

FC 17. What solution has the lesser concentration of water, thus "pulls" the water?
a. Hypertonic
b. Hypotonic
c. Aquatonic
d. Isotonic

18. If water is moving from the ECF to the ICF, the ICF is:
a. hypertonic.
b. hypotonic.
c. aquatonic.
d. isotonic.

19. When molecules diffuse across a membrane, they are said to:
a. penetrate the membrane.
b. pass the membrane.
c. permeate the membrane.
d. cross the membrane.

20. If a membrane only allows some molecules to diffuse through it but not others, it is said to be what?
a. Porous
b. Selectively permeable
c. Aqueous
d. Impermeable

FC 21. If you have a glass of soda, what is the solvent?
a. Fructose
b. Glucose
c. Water
d. Artificial colors and flavors

22. Which of the following are special water-conducting channels involved in osmosis?
a. Gate channels
b. Osmotic pores
c. Osmotic channels
d. Aquaporins

23. Gated channels are examples of:
a. carrier-mediated passive transport.
b. channel-mediated passive transport.
c. facilitated active transport.
d. osmotic channels.

24. Which of the following forms of transport changes shape to release the bound molecule?
a. Carrier-mediated passive transport
b. Channel-mediated passive transport
c. Facilitated active transport
d. Osmotic channels

25. A white blood cell ingesting a bacteria is an example of:
a. exocytosis.
b. phagocytosis.
c. pinocytosis.
d. osmosis.

FC ***Short answer***

26. Name the two main forms of transport by vesicles.

27. Name the two forms of endocytosis.

28. What are the three terms used to describe osmotic concentrations?

29. What are the two major forms of facilitated diffusion?

CELL METABOLISM

Fill in the blanks with the correct answer.

30. ____________________ act in an allosteric fashion on proenzymes, thus activating them.

31. Large changes in temperature, pH, and many other factors may destroy the configuration of an enzyme in a process called ____________________.

FC 32. ____________________ refers to all chemical reactions and is the basis for all human functions.

33. Enzymes are ____________________ and often their molecules contain a nonprotein part called a(n) ____________________.

FC 34. ____________________ reactions usually require the input of energy, while ____________________ pathways usually release energy.

Multiple choice – select the best answer.

35. An enzyme synthesized in an inactive form is called a:
   a. proenzyme.
   b. kinase.
   c. coenzyme.
   d. catalyst.

FC 36. The metabolic pathway that involves the breakdown of material is called:
   a. catalytic.
   b. catabolic.
   c. enzymatic.
   d. anabolic.

FC 37. Functional protein enzymes are called:
   a. cofactors.
   b. coenzymes.
   c. catalysts.
   d. proenzymes.

38. Organic nonprotein cofactors are called:
   a. cofactors.
   b. coenzymes.
   c. catalysts.
   d. proenzymes.

39. A molecule that alters enzyme function by changing its shape is called a(n):
   a. catalyst.
   b. cofactor.
   c. allosteric effector.
   d. end-product inhibitor.

40. In the example of enterokinase changing trypsinogen into trypsin, trypsinogen is an example of what?
   a. Proenzyme
   b. Kinase
   c. Cofactor
   d. Coenzyme

CELL RESPIRATION

Fill in the blanks with the correct answer.

41. ____________________ is the continuous process of breaking down carbohydrates, proteins, and fats into carbon dioxide and water.

42. ____________________ metabolism does not require oxygen and actually requires energy input to begin, whereas ____________________ pathways require oxygen to be completed.

43. ____________________ is broken down in the ____________________, forming pyruvic acid. Without oxygen, pyruvic acid is converted to ____________________. If oxygen is available, pyruvic acid is transferred to ____________________ pathways called the ____________________ cycle and the ____________________ system.

44. ____________________ is the metabolic process that breaks down complex organic molecules into smaller ones.

45. Cellular respiration is divided into three smaller pathways: ____________________, ____________________, and ____________________.

FC *Matching – match each description with the correct term.*

	Term		Definition
46. ______	Anabolism	a.	Portion of respiration that does NOT require oxygen
47. ______	Electron transport system	b.	Governed by a series of enzymes that are found in the inner chamber of the mitochondrion
48. ______	Citric acid cycle	c.	Protons moving down their concentration gradient in the inner mitochondrial membrane
49. ______	Aerobic pathway	d.	Occurs in cytosol and literally means "breaking glucose"
50. ______	Anaerobic pathway	e.	Protein synthesis
51. ______	Glycolysis	f.	Citric acid cycle and electron transport system

Multiple choice – select the best answer.

FC 52. The waste products of the citric acid cycle and electron transport system are:
a. lactic acid and carbon dioxide.
b. water and carbon dioxide.
c. pyruvic acid and water.
d. citric acid and carbon dioxide.

53. What is traveling down a concentration gradient and is responsible for most of the energy produced from the electron transport system?
a. Electrons
b. Protons
c. Neutrons
d. Citric acid

FC 54. Glycolysis:
a. takes place only in the cytosol.
b. takes place in the cytosol and mitochondrion.
c. requires oxygen.
d. converts pyruvic acid to an acetyl molecule.

55. This is a continuous process of breaking down carbohydrates, proteins, and fats into carbon dioxide and water.
a. Glycolysis
b. Cellular respiration
c. Anaerobic respiration
d. Anabolic respiration

56. Without oxygen, pyruvic acid will continue on an anaerobic pathway to form a molecule of:
a. citric acid.
b. acetic acid.
c. oxaloacetic acid.
d. lactic acid.

57. With oxygen, lactic acid will be converted:
a. to carbon dioxide and water.
b. to ATP.
c. back to pyruvic acid or glucose.
d. to acetic acid and ATP.

STUDY TIP

Think about "Aerobics." The term "Aerobics," as in the exercise, is taken from science; however, is it really "aerobic"? In general, if you are out of breath or unable to carry on a conversation, you are in anaerobic metabolism and not burning fat but using up your glycogen stores and producing lactic acid. I have observed many and even participated in a few Aerobics classes and depending on your physical conditioning, you may not be doing "aerobics" but "anaerobics." Interesting!

CHAPTER 5

Cell Growth and Reproduction

HOW TO APPROACH CELL GROWTH AND REPRODUCTION

Mitosis and meiosis are the two different types of cellular reproduction. Each has different stages with different end results. Know all the stages of each, what is happening at each stage, and where they are happening. See the study tip below for stages of mitosis. The most important thing to understand with mitosis is what is involved in copying our genetic code. To understand this, you must first memorize all the components of DNA and what they do. Once you have all the definitions put to memory, you can proceed to understanding how our genetic code is copied. Try drawing a picture of a cell with the nucleus. Label the important components and steps in replication. Next time you are watching TV, pay attention when they talk about "DNA, chromosomes, or genes" and see if it makes more sense or if they completely got it wrong. Hollywood frequently gets it wrong, but even the news does not always get it right.

DNA/RNA

Fill in the blanks with the correct answer.

1. ____________________ is the study based on DNA and related molecules.

2. ____________________ is a large molecule that is an extremely long polymer made up of many smaller molecules called *nucleotides*.

3. The ____________________ of DNA are held together by hydrogen bonds.

FC 4. A gene is a segment of DNA that codes for a specific ____________________.

5. ____________________ carries amino acids to the ribosome for placement in a prescribed sequence.

6. A(n) ____________________ is many ribosomes positioned in a "train" along a single mRNA molecule.

FC *Matching – match each description with the correct term.*

	Term	Definition
7. ______	Nucleotides	a. Segment of three nucleotides
8. ______	Gene	b. Production of an mRNA molecule
9. ______	Codon	c. Transfer of amino acids to the ribosome for placement in the prescribed sequence
10. ______	mRNA	d. Complete set of proteins synthesized by a cell
11. ______	Transcription	e. Entire set of genes in a cell
12. ______	Translation	f. Consist of a base, phosphate, and sugar
13. ______	Proteome	g. Copy or transcript of a gene
14. ______	Genome	h. Has amino acid attached to it
15. ______	tRNA	i. A segment of DNA that codes for a specific protein

Multiple choice – select the best answer.

16. *Obligatory base pairing* means that adenine will always pair with:
   a. cytosine.
   b. guanine.
   c. thymine.
   d. adenine.

17. DNA is made of smaller molecules called:
   a. genes.
   b. bases.
   c. genomes.
   d. nucleotides.

18. Bases of the DNA molecule are held together by what?
   a. Hydrogen bonds
   b. tRNA
   c. mRNA
   d. Nucleotides

19. The transcript code is read in segments of three nucleotides called a:
   a. codon.
   b. gene.
   c. genome.
   d. interon.

20. Which of the following is the best description of transcription?
   a. Takes place in the nucleus
   b. Production of a tRNA molecule
   c. Brings amino acids to the ribosome
   d. Creates an mRNA/ribosomal complex

21. Which of the following only happens in the cytoplasm?
   a. Translation
   b. Transcription
   c. Complementary base pairing
   d. Removing interons by spliceosomes

22. Amino acids are connected to which of the following?
   a. DNA
   b. mRNA
   c. tRNA
   d. proteomes

23. Translation produces:
   a. polypeptides.
   b. lipids.
   c. RNA.
   d. sugar molecules.

24. Nucleotides are made of what?
   a. DNA and RNA
   b. RNA only
   c. Four bases
   d. Phosphate, sugar, and base

FC *Short answer*

25. List the four nitrogenous bases of DNA.

26. List the four nitrogenous bases of RNA.

27. List the types of RNA.

Labeling – label the following diagram.

28. Protein synthesis

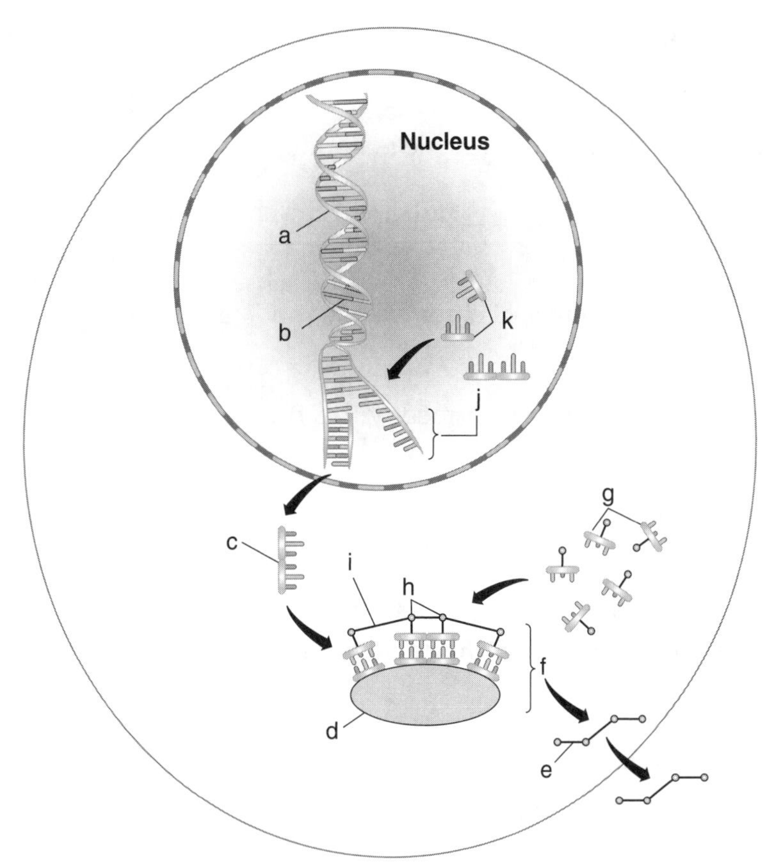

a. ____________________

b. ____________________

c. ____________________

d. ____________________

e. ____________________

f. ____________________

g. ____________________

h. ____________________

i. ____________________

j. ____________________

k. ____________________

CELL GROWTH AND REPRODUCTION

STUDY TIP

FC How to remember the six stages of cellular reproduction in order:

I	*Interphase*
Put	*Prophase*
My	*Metaphase*
Ass	*Anaphase*
Through	*Telophase*
College	*Cytokinesis*

For the four phases of mitosis, just remember PMAT or "Pee-MAT."

Fill in the blanks with the correct answer.

FC 29. ____________________ produces essentially a clone cell, whereas ____________________ produces a cell with one-half the genetic material.

30. In anaphase, the ____________________ pull apart the chromosome to form two single chromatids.

31. Gametes contain half of the number of chromosomes and are called ____________________ cells.

32. After fertilization, a new individual is formed with a full complement of genetic material called a(n) ____________________.

33. Normal body cells containing (give the #) ____________________ chromosomes are called *diploid cells.*

FC ***Matching – match each description with the correct term.***

	Term		Definition
34. ______	Mitosis	a.	"First phase"
35. ______	Meiosis	b.	"End phase"
36. ______	Interphase	c.	Replication of DNA
37. ______	Cytokinesis	d.	Creation of two new identical nuclei
38. ______	Prophase	e.	Chromosomes line up
39. ______	Anaphase	f.	Chromosome splits
40. ______	Telophase	g.	Creating gametes
41. ______	Metaphase	h.	Pinching cell membrane in half

Multiple choice – select the best answer.

42. Actual copying of DNA occurs during what phase?
    a. Cytokinesis
    b. Prophase
    c. Metaphase
    d. Interphase

43. Chromatids are held together by:
    a. chromosomes.
    b. centromeres.
    c. spindle fibers.
    d. centrioles.

44. A zygote is created by:
    a. meiosis.
    b. combining haploid cells.
    c. mitosis.
    d. combining diploid cells.

45. A chromosome consists of what?
    a. Sister chromatids
    b. One long DNA strand forming an X
    c. A diploid pair
    d. A haploid pair

46. Which of the following best describes mitosis?
    a. Cytoskeleton pinches the cell membrane in, creating two new cells
    b. Involves two haploid cells
    c. Results in two diploid cells
    d. Produces gametes

47. The specific period of interphase where DNA is replicated is called:
    a. S-phase.
    b. G_1-phase.
    c. G_2- phase.
    d. M-phase.

48. In what phase of mitosis does the nuclear envelope disappear?
    a. Telophase
    b. Metaphase
    c. Anaphase
    d. Prophase

49. Two new nuclear envelopes begin to form around the separated chromosomes, marking the beginning of:
    a. prophase.
    b. metaphase.
    c. anaphase.
    d. telophase.

50. Haploid cells are produced:
    a. by gametes.
    b. through meiosis.
    c. during cytokinesis.
    d. by fertilization.

51. During which phase does the chromosome split into chromatids?
    a. Prophase
    b. Anaphase
    c. Metaphase
    d. Telophase

52. During which phase does each chromatid attach to a spindle fiber?
    a. Cytokinesis
    b. Interphase
    c. Metaphase
    d. Anaphase

53. Which of the following contain 23 homologous pairs of chromosomes?
    a. Gametes
    b. Meiotic cells
    c. Diploid cells
    d. Zygotes

Short answer

FC 54. Identify the six phases of the cell's growth cycle.

FC 55. Identify the four phases of mitosis.

56. What are the three phases of interphase in order?

Labeling – Identify the following phases of cellular reproduction.

57. Events of mitosis

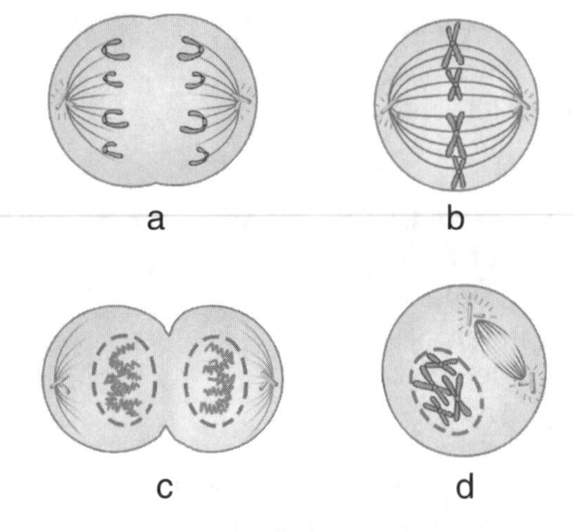

a. ______________________

b. ______________________

c. ______________________

d. ______________________

CHAPTER 6

Tissues and Their Functions

HOW TO APPROACH THE TISSUES AND THEIR FUNCTIONS

All the tissues have similarities but the key is to focus on what is unique or different about each tissue. Epithelial always faces an environment (COVERS THINGS) on one side and the basement membrane on the other. Very similar to carpet or wallpaper. It also is avascular yet regenerates quickly. Muscle MOVES things (connective tissue) by SHORTENING (contracting). Neural tissue is either a neuron or neuroglia. Neurons send, interpret, and receive INFORMATION. Neuroglia do all the other work like protect, insulate, and so forth. See the study tip below. Connective tissue has a MATRIX. *Matrix* is what surrounds every connective tissue cell. The matrix of blood is plasma; the matrix of bone is "bone" which is mostly calcium, magnesium, and other minerals. Make sure you know all of the different types of connective tissue and where to find each of them. This goes for all the tissues. You need to know all four tissue types, all of the cell types under each tissue type, what each of the cell types does, where it is found, and to be able to recognize them visually.

STUDY TIP

As you go through the text and look at the various tissue types, try to make note of unique visual characteristics that will help you remember that particular tissue type. For example, adipose looks like Styrofoam, bone looks like bunch of targets, reticular has NO straight lines (fibers), dense is a bunch of straight lines all going in the same direction. It is best if you make the associations yourself. Hopefully, these examples will get you started.

PRINCIPAL TYPES OF TISSUE AND EXTRACELLULAR MATRIX

Fill in the blanks with the correct answers.

FC 1. ____________________ are groups of similar cells performing a common function.

2. A complex mix of fluids and proteins is usually referred to as ____________________.

3. The study of tissues is called ____________________.

4. ____________________ are large molecules constructed of a protein backbone linked to carbohydrates such as sugars.

5. Structural proteins are often connected to cells by other modified proteins called ____________________.

Matching – match each term with the correct description (answers may be used more than once).

	Definition	**Term**
6. ______	Large quantities of matrix	a. Epithelial tissue
7. ______	Secretion, excretion, and absorption	b. Connective tissue
8. ______	Movement	c. Muscle tissue
9. ______	Covers and protects	d. Nervous tissue
10. ______	Communication	
11. ______	Glands	
12. ______	Supports	
13. ______	Contractile units	
14. ______	Electrical signals	
15. ______	Lines body cavities	

Short answer

FC 16. List the four principal types of tissue.

17. Identify the three primary germ layers.

18. Identify two common structural proteins.

Labeling – Identify the following cell and tissue types (example – "Skeletal, muscle" or "Simple squamous, epithelium"). (Answer blanks on next page.)

19. Tissue types

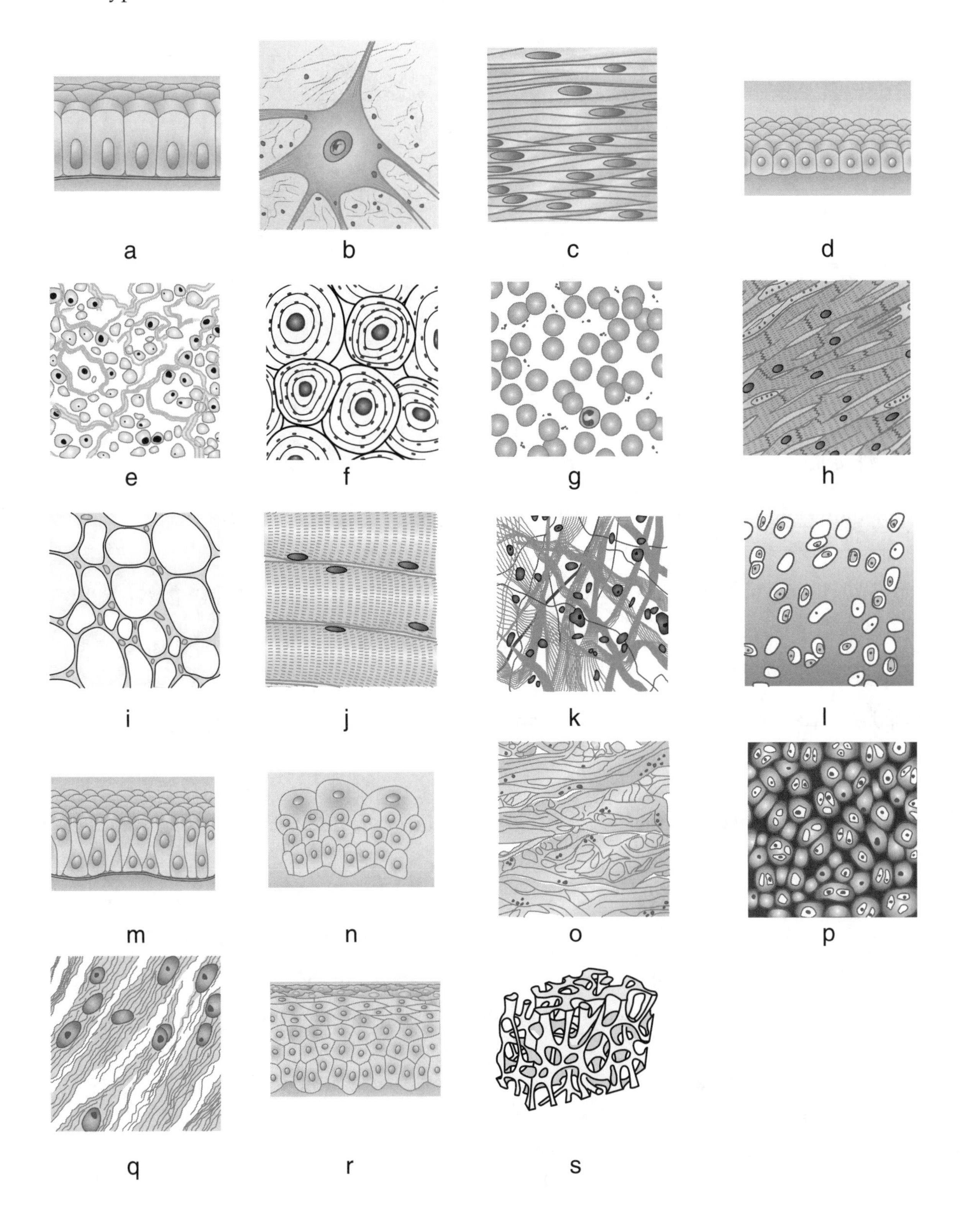

a. ______________________

b. ______________________

c. ______________________

d. ______________________

e. ______________________

f. ______________________

g. ______________________

h. ______________________

i. ______________________

j. ______________________

k. ______________________

l. ______________________

m. ______________________

n. ______________________

o. ______________________

p. ______________________

q. ______________________

r. ______________________

s. ______________________

EPITHELIAL TISSUE

Fill in the blanks with the correct answers.

20. Epithelium is divided into two general types: ______________ and ______________.

21. Some columnar epithelial cells are modified into ______________ that produce mucus.

22. ______________ are microscopic cellular projections that increase the surface area on columnar cells for increased absorption.

23. Based on structural classification of exocrine glands, ducts can be either ______________ or ______________.

FC ***Matching – identify the epithelial type that would be in the following locations (types may be used more than once).***

	Locations	**Epithelial type**
24. ______	Tubules of the kidney	a. Simple squamous
25. ______	Possess cilia or microvilli	b. Simple cuboidal
26. ______	Air passages of the respiratory system	c. Simple columnar
27. ______	"Skin"	d. Pseudostratified columnar
28. ______	Lining blood and lymphatic vessels	e. Stratified squamous
29. ______	Urinary bladder	f. Stratified cuboidal
30. ______	Pharynx and parts of the epiglottis	g. Transitional
31. ______	Many types of glands and their ducts	
32. ______	Stomach and intestine	
33. ______	Air sacs of the lungs	

Multiple choice – select the best answer.

FC 34. What connects the epithelial tissue to the connective tissue below?
a. Dermis
b. Basement membrane
c. Membranous epithelium
d. Subepithelial tissue

35. Which of the following best describes epithelial tissue?
a. Avascular
b. Extracellular matrix
c. Rich blood supply
d. Consists predominantly of fibers

36. Nonkeratinized stratified squamous epithelium is located in all the following EXCEPT the:
a. vagina.
b. mouth.
c. trachea.
d. esophagus.

37. Milk is produced by which of the following type of glands?
a. Holocrine
b. Apocrine
c. Merocrine
d. Lactocrine

38. This type of gland collects its products inside the entire cell and ruptures completely to release it.
a. Holocrine
b. Apocrine
c. Merocrine
d. Lactocrine

FC 39. What type of gland releases its products directly into the bloodstream?
a. Endocrine
b. Merocrine
c. Holocrine
d. Exocrine

40. Keratin is a special hard protein found in which type of epithelium?
a. Transitional
b. Stratified cuboidal
c. Simple columnar
d. Stratified squamous

FC 41. Which of the following types of glands cause no damage to the cell while releasing their product?
a. Holocrine
b. Merocrine
c. Mammary
d. Sebaceous

FC ***Short answer***

42. List five functions of epithelial tissue.

43. List the four groups of epithelium based on shape.

44. List the three type of epithelial tissue based on layers of cells.

45. Identify the three types of glands based on functional classification.

Labeling – Identify the following types of glands.

46. Three types of exocrine glands

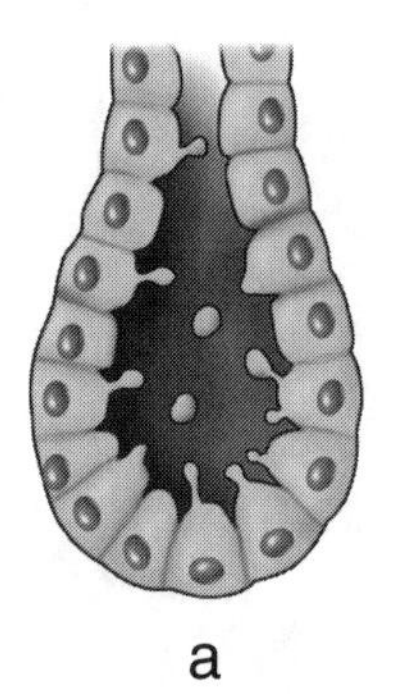

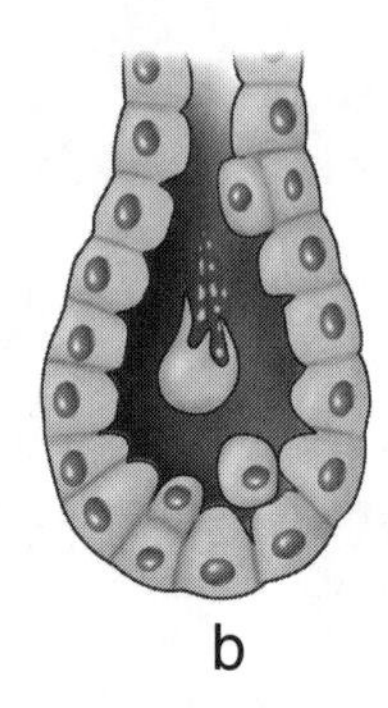

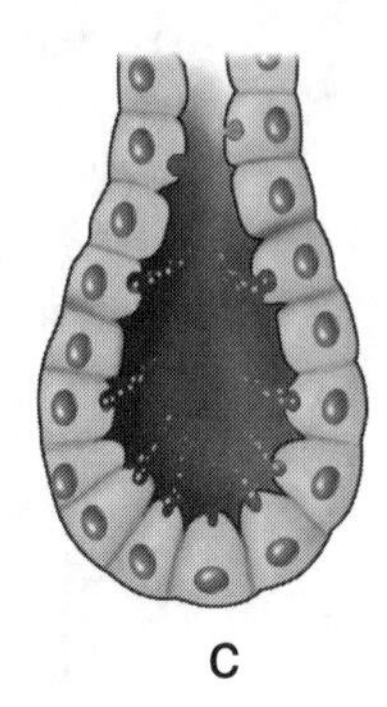

a. ______________________

b. ______________________

c. ______________________

CONNECTIVE TISSUE

Fill in the blanks with the correct answers.

FC 47. ____________________ connect muscle to bone and ____________________ connect bone to bone.

FC 48. Dense fibrous tissue is divided into ____________________ and ____________________ categories, depending on the arrangement of the fibers.

49. Bone cells are called ____________________, bone-forming cells are called ____________________, and bone-digesting cells are called ____________________.

50. Tiny canals called ____________________ ("little canals") connect lacunae with each other and with the blood vessels from the ____________________.

Matching – match each description or location in the body with the correct term.

	Term		Definition
51. ______	Areolar	a.	Ear
52. ______	Adipose	b.	Spleen, lymph nodes, and bone marrow
53. ______	Reticular	c.	Capsules surrounding organs and muscles
54. ______	Dense regular	d.	Ends of long bones
55. ______	Dense irregular	e.	Intervertebral disks
56. ______	Osteon	f.	Subcutaneous tissue, under the skin
57. ______	Hyaline cartilage	g.	Tendons and ligaments
58. ______	Fibrocartilage	h.	Basic organizational unit of compact bone
59. ______	Elastic cartilage	i.	"Fat"

Multiple choice – select the best answer.

60. What consists predominantly of fluid, varying numbers and types of fibers, and ground substances?
    a. Adipose
    b. Extracellular matrix
    c. Collagen
    d. Lamellae

FC 61. Which fiber type is extremely abundant and extremely strong?
    a. Collagenous
    b. Reticular
    c. Elastic
    d. Dense

FC 62. Which of the following cell types is the most numerous in loose connective tissue and produces a gel-like ground substance?
    a. Macrophages
    b. Mast cells
    c. Areolar cells
    d. Fibroblasts

FC 63. Which of the following cells produce histamine, heparin, leukotrienes, and prostaglandins?
    a. Macrophages
    b. Mast cells
    c. Fibroblasts
    d. Histocytes

64. Dense fibrous connective tissue consists mostly of which of the following fiber types?
    a. Collagenous fibers
    b. Dense fibers
    c. Reticular fibers
    d. Elastic fibers

FC 65. Osteocytes are located in small spaces called:
    a. lamellae.
    b. canaliculi.
    c. osteons.
    d. lacunae.

FC 66. What are bone "building" cells?
    a. Osteoblasts
    b. Osteoclasts
    c. Osteocytes
    d. Blastocytes

67. Which of the following connective tissues contains neither ground substance nor fibers?
    a. Elastic cartilage
    b. Areolar tissue
    c. Adipose tissue
    d. Blood

FC 68. White blood cells are called:
    a. erythrocytes.
    b. leukocytes.
    c. thrombocytes.
    d. osteocytes.

FC *Short answer*

69. List the three fibers that comprise the extracellular matrix.

70. Identify the three types of cartilage.

71. List three components of an osteon.

72. What are the three cell types of blood?

MUSCLE AND NERVOUS TISSUE, TISSUE REPAIR, AND BODY MEMBRANES

STUDY TIP

FC To remember the difference between the functions of smooth and skeletal muscle, just think about the fable of the tortoise and the hare. Smooth muscle is the tortoise; slow, continuous, and never stops. Skeletal muscle is the hare; fast, powerful, dynamic, yet tires quickly. Cardiac muscle is the best of both without their weaknesses. What would be the best representation of a battery that never stops, the tortoise or the hare?

Fill in the blanks with the correct answers.

73. ____________________ tissue contains cells that are specialized for contractility.

FC 74. The ____________________ membrane is the membrane that lines cavities and the ____________________ membrane covers the organs in the cavities.

FC 75. The ____________________ surround the lungs and line the thoracic cavity, and the ____________________ covers the abdominal viscera and lines the abdominal cavity.

FC 76. The ____________________ is the serous membrane that surrounds the heart.

77. ____________________ tissue deals with sending, processing, and interpreting information.

Matching – match each description, location, or word with the correct term.

Term

78. ______ Skeletal muscle
79. ______ Smooth muscle
80. ______ Cardiac
81. ______ Neuroglia
82. ______ Soma
83. ______ Mucous membrane
84. ______ Serous membrane
85. ______ Axon
86. ______ Cutaneous membrane
87. ______ Regeneration
88. ______ Synovial membrane

Definition

a. Line cavities that are not open to the external environment
b. "Skin"
c. Transmits nerve impulses away from soma
d. Intercalated discs
e. Connective tissue membrane
f. Involuntary
g. Striated and multinucleated
h. Removing dead cells while other cells divide to replace dead cells
i. Connect and protect neurons
j. Neuron cell body
k. Located in the respiratory, digestive, urinary, and reproductive tracts

Multiple choice – select the best answer.

89. Smooth muscle would be located in all of the following locations EXCEPT:
    a. stomach.
    b. intestines.
    c. blood vessels.
    d. hand.

FC 90. Intercalated discs are associated with which of the following cell types?
    a. Neurons
    b. Neuroglia
    c. Astrocytes
    d. Cardiac

FC 91. What carries nerve signals toward the cell body?
    a. Soma
    b. Axons
    c. Dendrites
    d. Microglia

92. Which of the following types of tissue do NOT regenerate?
    a. Connective and muscle
    b. Nervous and epithelial
    c. Connective, muscle, and nervous
    d. Muscle and nervous

93. This membrane contains many sweat and oil glands.
    a. Cutaneous
    b. Serous
    c. Mucous
    d. Glandular

94. What are the small synovial sacs located between some moving body parts called?
    a. Lamina propria
    b. Synovial joints
    c. Synovial compartments
    d. Bursae

95. Which membrane is the portion that lines the walls of cavities?
    a. Visceral
    b. Parietal
    c. Peritoneal
    d. Pleural

96. Fibrous connective tissue underlying the epithelium in mucous membranes is called the:
    a. lamina propria.
    b. basement membrane.
    c. visceral peritoneum.
    d. subcutaneous sheet.

Short answer

FC 97. Identify the three types of muscle tissue.

FC 98. List the three general types of epithelial tissue membranes in the body.

99. Name three locations of smooth muscle.

Labeling – Identify the following membranes.

100. Types of body membranes

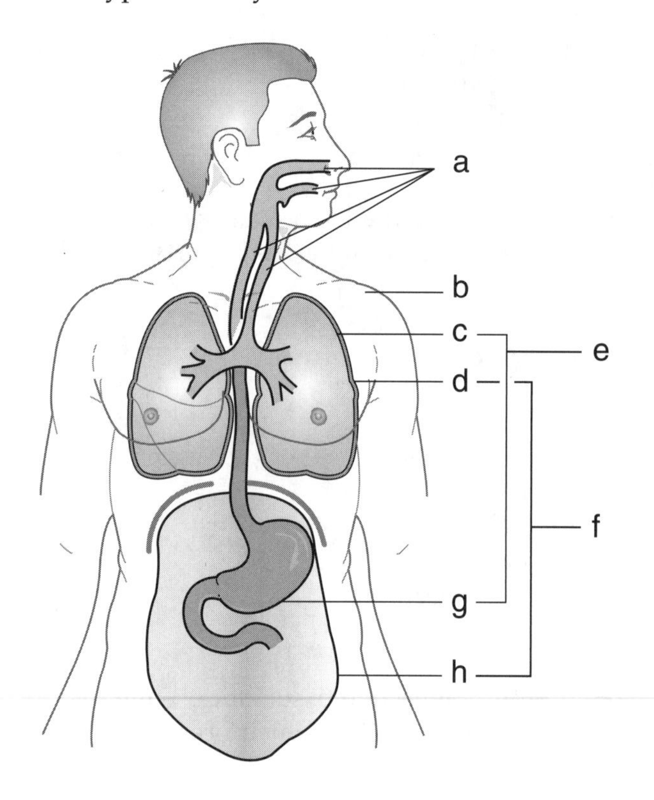

a. ______________________________

b. ______________________________

c. ______________________________

d. ______________________________

e. ______________________________

f. ______________________________

g. ______________________________

h. ______________________________

STUDY TIP

Think of the classic stereotype of a "nerd." A nerd is really good at "information or data" but really bad at most everything else such as cooking, dressing, self-defense, and dating. Now think about the neuron—same as the nerd. It's all about INFORMATION and nothing else. The glial cells do all those things for the neuron. Thus you can remember "nerdy neuron."

CHAPTER 7

Skin and Its Appendages

HOW TO APPROACH THE SKIN AND ITS APPENDAGES

Before starting to answer these questions, you should be able to name all the layers of the skin from superficial to profundus. See the study tip below for the five layers of the epidermis. Which structures are found in each layer of the skin and what is their function? For example, melanocytes are located in the stratum basale and spinosum and are responsible for producing the pigment melanin, and the epidermis is epithelial tissue, whereas the dermis is connective tissue. Think back to the previous chapter and what is unique to these two types of tissues. You can then identify what is found or not found in each of these layers of the skin. For example, epithelium does not contain blood vessels; thus, hair must originate in the dermis. The dermis has only two layers and consists of connective tissue also containing two other types of tissue: muscle and nervous. It is easier to first learn the structures (layers of the skin), and then their functions. Be sure you are comfortable with directions superficial and profundus.

STRUCTURE OF THE SKIN

STUDY TIP

FC How to remember the layers of the epidermis, superficial to profundus.

1. Stratum **C**orneum (core-knee-um) = **C**harlie
2. Stratum **L**ucidum (lu-sye-dum) = **L**ikes
3. Stratum **G**ranulosum (gran-you-low'sum) = **G**rabbing
4. Stratum **S**pinosum (spy-know'sum) = **S**hort
5. Stratum **B**asale / Germinativum (bah-say'lee) = **B**eards

We will refer to "Charlie" throughout the book.

Fill in the blanks with the correct answers.

1. ____________________ disks or ____________________ cells serve as receptors for light touch.

2. ____________________ may form in areas of constant friction or irritation.

3. Between the dermis and epidermis lies the ____________________ or DEJ.

FC 4. The ____________________ is sometimes called the *subcutaneous layer*.

5. The majority of epidermal cells are filled with a tough, fibrous protein called ____________________.

FC ***Matching – match each description with the correct term.***

	Term		Definition
6. ______	Papillary layer	a.	"Fingerprints"
7. ______	Reticular layer	b.	Produces all cells of the epidermis
8. ______	Stratum corneum	c.	Also called *tactile epithelial cell*
9. ______	Merkel cells	d.	Most superficial layer of skin
10. ______	Keratinocytes	e.	Most superficial layer of dermis
11. ______	Stratum basale	f.	Connective tissue layer of skin
12. ______	Melanocytes	g.	Most profundus layer of dermis
13. ______	Stratum lucidum	h.	90% of epidermal cells
14. ______	Dermis	i.	Give skin its color
15. ______	Friction ridges	j.	Thickest on soles and palms

Multiple choice – select the best answer.

FC 16. Which of the following is the most superficial layer of the skin?
 a. Dermis
 b. Papillary layer
 c. Stratum basale
 d. Stratum corneum

17. Friction ridges are produced by:
 a. dermal papillae.
 b. corneal ridges.
 c. reticular layer.
 d. keratinocytes.

FC 18. Which of the following absorbs ultraviolet radiation protecting the underlying layer?
 a. Keratinocytes
 b. Melanocytes
 c. Stratum corneum
 d. Papillary layer

FC 19. The tough, fibrous protein found in most epidermal cells is called:
 a. melanin.
 b. strata.
 c. keratin.
 d. collagen.

20. These cells identify invading bacteria and other invaders of the skin and are located in the epidermis.
 a. Merkel discs
 b. Leukocytes
 c. Merkel cells
 d. Dendritic cells

21. What is the deepest layer of the skin?
 a. Stratum basale
 b. Papillary layer
 c. Hypodermis
 d. Reticular layer

22. What helps regulate the regeneration and repair of the epidermis?
    a. Epidermal growth factor
    b. Keratinocytes
    c. Dermoepidermal cells
    d. Basement membrane

23. What forms a connection between the skin and the underlying structures of the body?
    a. Reticular layer
    b. Dermis
    c. Hypodermis
    d. Hypoepidermis

24. The dermoepidermal junction (DEJ), is mostly composed of the:
    a. basement membrane.
    b. stratum basale.
    c. papillary layer.
    d. dermoepidermal cells.

25. Which layer of the skin protects the body from radiation, chemicals, abrasions, cuts, and bacteria?
    a. Stratum lucidum
    b. Stratum granulosum
    c. Stratum germinativum
    d. Stratum corneum

26. Which of the epidermal layers is thickest on the soles of the feet and palms of the hands?
    a. Stratum lucidum
    b. Stratum granulosum
    c. Stratum germinativum
    d. Stratum corneum

27. This layer of the skin serves as an area of attachment for numerous skeletal and smooth muscle fibers.
    a. Papillary layer
    b. Hypodermis
    c. Subcutaneous layer
    d. Reticular layer

28. The majority of the specialized sensory receptors are located in the:
    a. epidermis.
    b. epidermal junction.
    c. dermis.
    d. hypodermis.

Labeling – label the following diagram.

29. Layers of the skin

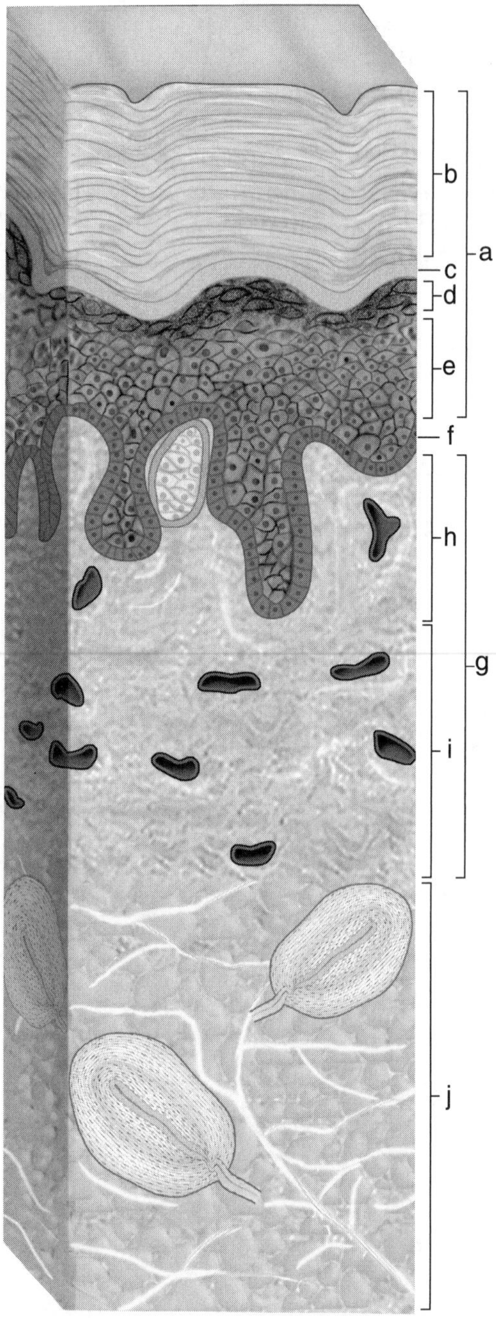

a. ____________________

b. ____________________

c. ____________________

d. ____________________

e. ____________________

f. ____________________

g. ____________________

h. ____________________

i. ____________________

j. ____________________

Short answer

30. Identify the two layers of the skin, profundus to superficial.

FC 31. Identify the five layers of the epidermis, superficial to profundus.

FC 32. List three epidermal cell types.

SKIN COLOR AND FUNCTIONS

Fill in the blanks with the correct answers.

33. The main determinant of our skin color is the pigment ____________________.

FC 34. ____________________ causes melanocytes to increase melanin production.

35. Skin that appears bluish due to large proportion of unoxygenated hemoglobin is called ____________________.

Multiple choice – select the best answer.

36. Melanosomes are:
    a. skin cells that produce melanocytes.
    b. dermal cells that produce melanin.
    c. tiny pigmented granules.
    d. the pigment produced by melanocytes.

FC 37. Prolonged exposure to ultraviolet radiation (UV) causes:
    a. melanocytes to increase melanin production.
    b. melanosomes to increase melanin production.
    c. melanin production to stop.
    d. melanin to increase melanocyte production.

38. What can cause yellowish discoloration of the skin?
    a. Melanin, beta-carotene, and bile pigments
    b. Melanocytes and bile
    c. Beta-carotene only
    d. Bile pigments and beta-carotene

39. Which of the following is NOT a function of the skin?
    a. Protection
    b. Excretion
    c. Contraction
    d. Immunity

40. Skin that is said to look *cyanotic* means that it appears:
    a. bluish.
    b. pale.
    c. moist.
    d. red.

41. The oily residue found on the surface of the skin is called:
    a. epidermal film.
    b. surface film.
    c. epi-film.
    d. cutaneous film.

42. Which of the following requires ultraviolet light and chemical precursor to be produced?
    a. Vitamin A
    b. Vitamin B
    c. Vitamin C
    d. Vitamin D

Short answer

43. Identify the seven functions of the skin.

44. List four things that can readily be absorbed by the skin.

FC 45. Identify the five sensory receptors located in the skin.

FC 46. Identify four fat-soluble vitamins than can be absorbed by the skin.

FC 47. Identify three waste products that can be excreted by the skin.

APPENDAGES OF THE SKIN

Fill in the blanks with the correct answers.

48. The fetus is covered by extremely fine hair called ____________________.

49. Soon after birth, the remaining hair is called ____________________ hair.

50. ____________________ glands secrete ____________________, an oily substance, into each hair follicle.

FC ***Matching – match the description, term, or location with the appropriate gland (answers may be used more than once).***

	Definition	Gland
51. ______	Armpit, areola of breast, and around anus	a. Sweat or sudoriferous
52. ______	Acne	b. Apocrine
53. ______	Produce a transparent watery liquid	c. Sebaceous
54. ______	External ear canal	d. Ceruminous
55. ______	Secrete oil	
56. ______	Most numerous	
57. ______	Produce "odor"	
58. ______	Contain ammonia, uric acid, and urea	
59. ______	Specialized sweat glands	
60. ______	At least two for each hair	

Multiple choice – select the best answer.

61. Adult hair is called:
    a. lanugo.
    b. vellus hair.
    c. terminal hair.
    d. mature hair.

62. As long as cells of what part of the hair are alive, hair will regenerate?
    a. Germinal matrix
    b. Shaft
    c. Medulla
    d. Cortex

63. Nails are composed of what?
    a. Mostly collagen and other protein fibers
    b. Heavily keratinized epidermal cells
    c. Substance similar to cartilage
    d. Mostly dead, keratinized dermal cells

64. What contains the capillaries that nourish the germinal matrix?
    a. Dermal root
    b. Epithelial root
    c. Hair papilla
    d. Medulla

65. What is the "little moon" of the nail called?
    a. Free edge
    b. Bed
    c. Root
    d. Lunula

FC 66. Body odor is associated with what type of gland?
    a. Sebaceous
    b. Apocrine
    c. Holocrine
    d. Eccrine

67. The medical term for baldness is called:
    a. vitiligo.
    b. comedo.
    c. onycholysis.
    d. alopecia.

68. The inner core of a hair is called the what?
    a. Cortex
    b. Cuticle
    c. Follicular core
    d. Medulla

69. What is the other name for sweat glands?
    a. Sebaceous
    b. Endocrine
    c. Holocrine
    d. Sudoriferous

70. Ceruminous glands are located where on the body?
    a. Armpits
    b. Next to the anus
    c. Ears
    d. Nipples

71. Which gland is associated with "pimples" or acne?
    a. Ceruminous
    b. Apocrine
    c. Sebaceous
    d. Sweat

Labeling – label the following diagrams.

72. Structure of nails

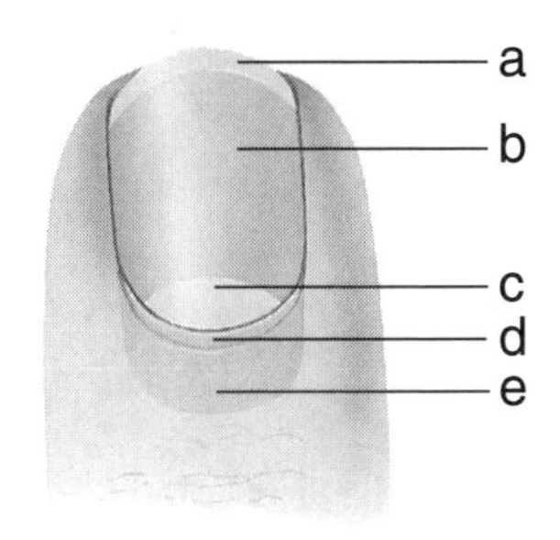

d b
e a

a. ______________________________

b. ______________________________

c. ______________________________

d. ______________________________

e. ______________________________

Short answer

73. Identify the five areas of the skin that are hairless.

FC 74. Identify the three skin glands.

75. Sebum contains what four things?

CHAPTER 8

Skeletal Tissues

HOW TO APPROACH SKELETAL TISSUES

This chapter is concerned with the types, structure, and formation of bone. All bone starts as compact bone, so make sure you are familiar with all the structures of compact bone or the osteon. Once you have a good understanding of the types of bone, move to the structures of the actual bone itself. The long bone is used as the example and divided into several parts. See the study tip below regarding this. The pictures and matching are particularly helpful with learning the structures of the long bone. You should be able to draw a quick long bone and label the parts with no need for artistic talent.

Once you have the above structures put to memory, focus on how you build and remodel bone. This will involve several different types of cells, hormones, and nonbiological materials. Most everyone knows one of the major materials is calcium, so focus on the things you don't know.

TYPES OF BONES

Fill in the blanks with the correct answers.

1. ____________________ bone is dense and solid, whereas ____________________ bone has open spaces partially filled by a network of fine, needlelike struts.

2. The thin, fibrous membrane lining the medullary cavity and spaces of spongy bone is the ____________________.

3. The ____________________ creates a hollow space throughout the diaphysis of a long bone.

Matching – match each description with the correct term.

	Term		Definition
4. ______	Long bones	a.	Hollow space through diaphysis
5. ______	Sesamoid bones	b.	Enlarged ends of long bone
6. ______	Irregular bones	c.	Uniquely shaped ends
7. ______	Short bones	d.	Surface of epiphyses
8. ______	Flat bones	e.	Patella
FC 9. ______	Epiphyses	f.	Shaft of long bones
FC 10. ______	Diaphyses	g.	Appear in coordinated groups
FC 11. ______	Periosteum	h.	Vertebral bones
FC 12. ______	Articular cartilage	i.	Sternum
FC 13. ______	Medullary cavity	j.	Covers bone exterior

Multiple choice – select the best answer.

14. The walls of the diaphysis are made of what?
    a. Periosteum
    b. Spongy bone
    c. Compact bone
    d. Endosteum

15. Muscles anchor to what part of the bone?
    a. Periosteum
    b. Diaphysis
    c. Epiphysis
    d. Endosteum

FC 16. Most bone growth takes place where?
    a. Ends of epiphyses
    b. Epiphyseal plate
    c. Periosteum
    d. Endosteum

17. The type of soft connective tissue found in the epiphyses that produces blood is called:
    a. yellow marrow.
    b. medullary cavity.
    c. red marrow.
    d. spongy bone.

18. Which of the following is found on the ends of the epiphyses?
    a. Diaphysis
    b. Periosteum
    c. Endosteum
    d. Articular cartilage

19. The other name for spongy bone is:
    a. cancellous.
    b. red marrow.
    c. yellow marrow.
    d. soft bone.

Labeling – label the following diagram.

20. Long bone

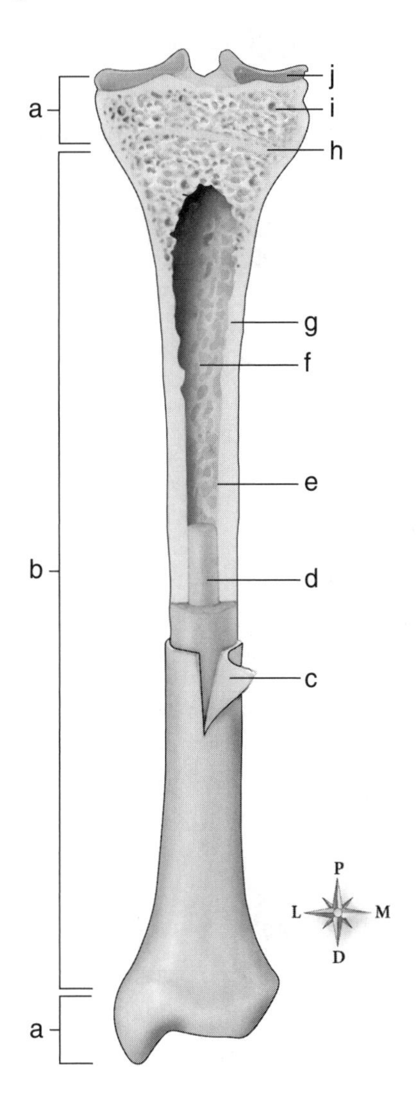

a. ______________________________

b. ______________________________

c. ______________________________

d. ______________________________

e. ______________________________

f. ______________________________

g. ______________________________

h. ______________________________

i. ______________________________

j. ______________________________

Short answer

21. Name five types of bones.

22. Name the six major parts of long bone.

BONE TISSUE AND MICROSCOPIC STRUCTURES OF BONE

Fill in the blanks with the correct answers.

23. The ____________________ is divided into two major chemical components: inorganic salts and organic salts.

24. During the complex ____________________ process, bone-forming cells secrete needlelike deposits of hydroxyapatite into the microscopic spaces between collagen fibers.

25. ____________________ and ____________________ are naturally occurring substances in the body and help cartilage remain smooth, compressible, and elastic and can be purchased over the counter as dietary supplements.

26. ____________________ are mature, nondividing bone cells that develop from osteoblasts.

FC *Matching – match each description with the correct term.*

	Term		Definition
27. ______	Osteocytes	a.	Allow osteocytes to communicate
28. ______	Osteoprogenitor cells	b.	Contains blood vessels, lymphatic vessels, and nerves
29. ______	Osteoclasts	c.	Unit of compact bone
30. ______	Osteoblasts	d.	Destroy bone
31. ______	Trabeculae	e.	Makes up spongy bone
32. ______	Central canal	f.	Build bone
33. ______	Canaliculi	g.	Layers of matrix
34. ______	Lacunae	h.	Stem cells that produce osteoblasts
35. ______	Lamellae	i.	Mature bone cells that develop from osteoblasts
36. ______	Osteons	j.	Where osteocytes reside

STUDY TIP

FC OsteoBlasts – BUILD, osteoClasts – CATABOLIZE, or simply remember BLASTs – BUILD. That must mean that clasts are the opposite. Also, you can remember **E**piphysis is the **END**s of the bone and **M**etaphysis is the **M**iddle. **CANAL**iculi are the **CANALS** in the **OSTEO**n and the cells are **OSTEO**cytes. Bone tissue is also called OSSEOUS tissue. You can find examples through this text to help remember things!

Multiple choice – select the best answer.

37. The rock-like crystals of hydroxyapatite are formed from:
    a. iron and magnesium.
    b. calcium and iron.
    c. calcium and phosphate.
    d. phosphate and iron.

38. What are "bone-eating" cells?
    a. Osteons
    b. Osteoblasts
    c. Osteocytes
    d. Osteoclasts

39. What organic substances are collectively called the "ground substance" of bone matrix?
    a. Collagen, polysaccharides, and protein
    b. Collagen and calcium
    c. Protein and polysaccharides
    d. Calcium, collagen, and polysaccharides

40. A single unit of compact bone is called a(n):
    a. lacunae.
    b. lamellae.
    c. osteon.
    d. osteocyte.

41. What type of bone cell "builds" bone?
    a. Osteocytes
    b. Osteoblasts
    c. Osteoclasts
    d. Osteoprogenitor cells

42. Needlelike branches of spongy bone are called:
    a. canaliculi.
    b. trabeculae.
    c. lacunae.
    d. canals.

43. Lamellae are:
    a. spaces filled with fluid where the bone cells live.
    b. cylinder-shaped layers of calcified matrix.
    c. vertical, cylinder-shaped units also called *haversian systems*.
    d. tiny canals allowing bone cells to communicate.

44. What are very fine canals that radiate in all directions from lacunae?
    a. Central canals
    b. Lateral canals
    c. Trabeculae
    d. Canaliculi

45. What are the cells located within the lacunae?
    a. Myeloid cells
    b. Osteocytes
    c. Osteoprogenitor cells
    d. Osteoblasts

46. Spongy bone is composed of needlelike branches called:
    a. lamellae.
    b. trabeculae.
    c. lamellar roots.
    d. osteon cylinders.

Labeling – label the following diagram.

47. Compact and cancellous bone in a long bone

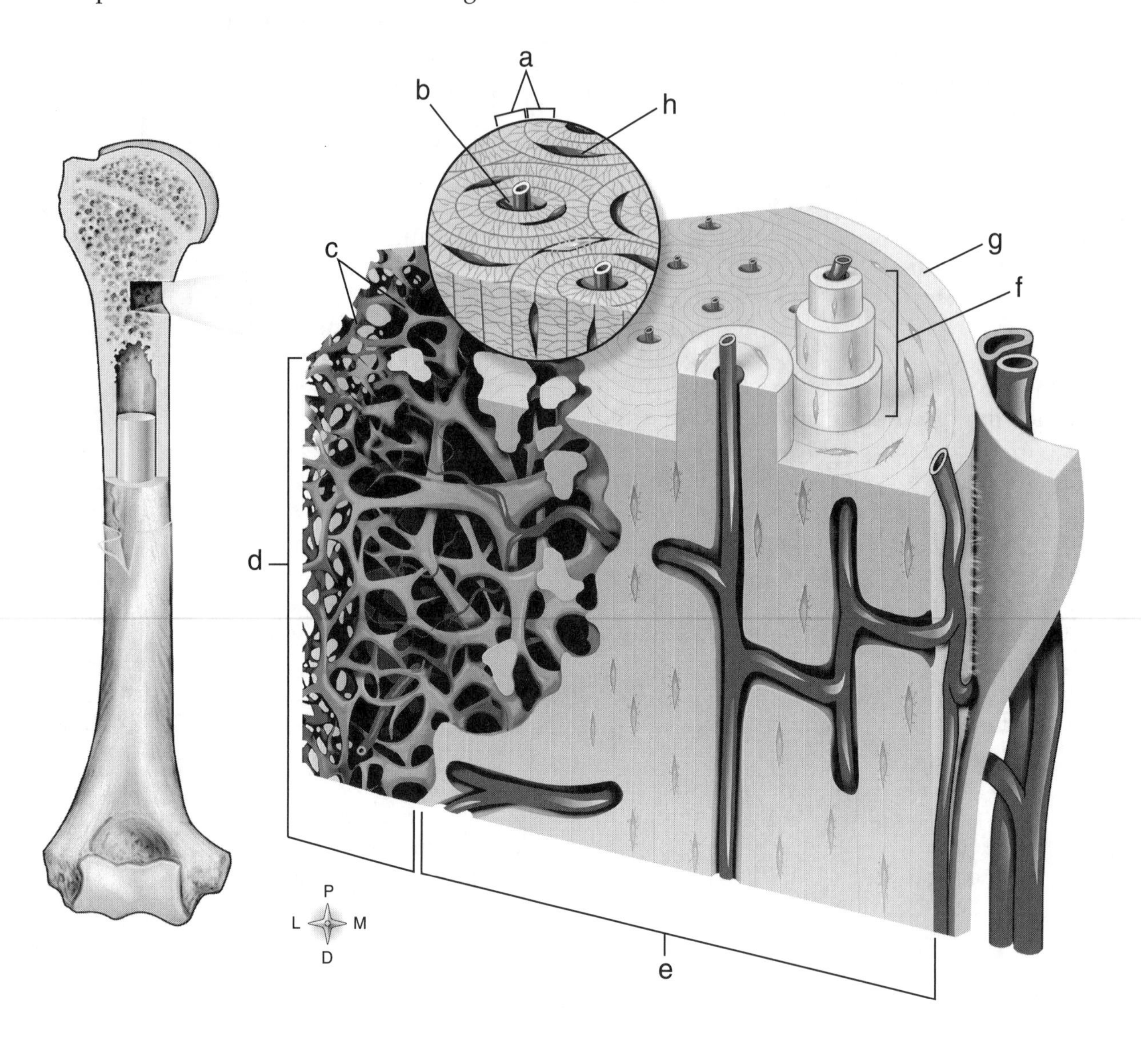

a. ______________________

b. ______________________

c. ______________________

d. ______________________

e. ______________________

f. ______________________

g. ______________________

h. ______________________

FC *Short answer*

48. List the four structures of an osteon.

49. Identify the three types of bone cells.

BONE MARROW, REGULATION, DEVELOPMENT, REMODELING, AND REPAIR, AND CARTILAGE

Fill in the blanks with the correct answers.

50. ____________________ enlarge the diameter of the medullary cavity.

51. ____________________ from the periosteum build new bone around the outside of the bone.

52. A ____________________ is a break in a bone or cartilage.

53. A broken bone creates a hemorrhage and a pool of blood at the point of injury called a ____________________. It quickly develops a mesh of fibrin filaments and transforms into ____________________ tissue. Soon cartilage forms and osteoblasts begin formation of a bony ____________________.

54. Cartilage lacks an abundant supply of blood vessels and nerves and is referred to as ____________________.

FC ***Matching – match each type of cartilage with the correct description or location (answers may be used more than once).***

	Definition	Term
55. ______	Symphysis pubis	a. Hyaline
56. ______	Most abundant type	b. Elastic
57. ______	External ear	c. Fibrocartilage
58. ______	Ends of long bones	
59. ______	Intervertebral disks	
60. ______	Looks "glassy" and somewhat transparent	
61. ______	Tip of nose	
62. ______	Most flexible	
63. ______	Forms auditory tubes	
64. ______	Very strong and rigid	

Multiple choice – select the best answer.

FC 65. The laying down of calcium salts by osteoblasts is called:
a. osteogenesis.
b. calcification.
c. ossification.
d. intramembranous ossification.

66. The location where bone growth proceeds along and down toward the diaphysis from each epiphysis is called the:
a. primary ossification center.
b. epiphyseal plate.
c. secondary ossification centers.
d. osteogenesis.

FC 67. Where are red blood cells produced?
a. Diaphysis
b. Red marrow
c. Yellow marrow
d. Osteocytes in epiphysis

FC 68. What induces osteoclasts to break down bone, releasing calcium into your blood?
a. Calcitonin
b. Vitamin D
c. Parathyroid hormone
d. Calcification hormone

69. Which of the following describes osteogenesis?
a. Combined action of osteoblasts and osteoclasts
b. Forming bone from cartilage
c. The laying down of calcium salts by osteoblasts
d. Demineralization of developing bones

70. Most flat bones are formed by a process called:
a. intramembranous ossification.
b. endochondral ossification.
c. primary ossification.
d. secondary ossification.

71. When a bone is broken, what binds the broken ends together?
a. Osseous hematoma
b. Fracture hematoma
c. Granulation tissue
d. Callus tissue

72. Which of the following is NOT a type of cartilage?
a. Hyaline
b. Membranous
c. Elastic
d. Fibrocartilage

73. Nutrients and gases diffuse slowly into cartilage from the fibrous covering called:
a. perichondrium.
b. membranous chondrocytes.
c. periosteum.
d. chondrostium.

74. Collagen fibers are most numerous in what type of cartilage?
a. Hyaline
b. Elastic
c. Fibrocartilage
d. Reticular

Labeling – Identify the following types of cartilage.

75. Types of cartilage

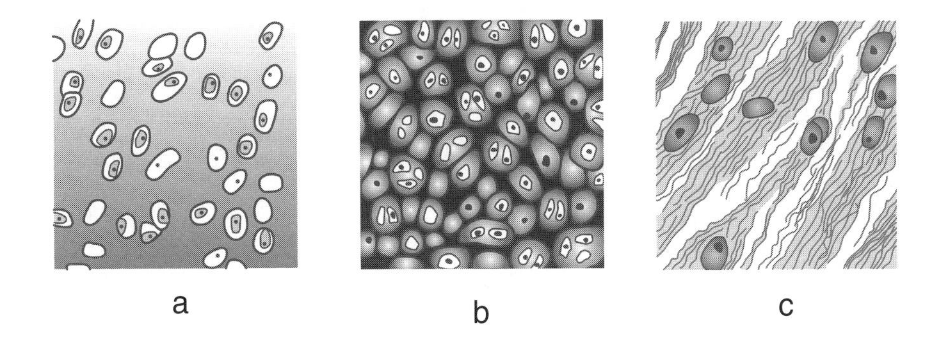

a. ______________________________

b. ______________________________

c. ______________________________

CHAPTER 9

Bones and Joints

HOW TO APPROACH THE BONES AND JOINTS

The primary focus of this chapter is anatomy or structure of the skeletal system. It is very similar to looking at a map and trying to memorize where the states are in relation to each other. The best way to approach this is by using pictures. You would not try to memorize the states by reading about their directions, you would look at a map over and over. Try this approach—flag the pages in your book and highlight all the structures (their names) pointed out on the picture that you have to know for your test. Then using your hand or a note card, cover the names and identify all of them that you have to know. You may have to highlight some of the lines going to the structures so you know which ones to try to name. When you are done naming them, uncover the names and see how many you got wrong. Keep doing this until you get them all right. I recommend ONLY doing this for about 15–20 minutes at a time. You can do as little as 5 minutes, but do it several times a day. It's all about the repetition, not the quantity. The more times you do this, the more you will remember. Two hours of this would be a complete waste of time. Flash cards are not recommended for this type of memorization; however, they are useful for questions like the short answers in this chapter. Use your text and the pictures. The matching section containing the terms like *tuberosity, epicondyle, foramen,* and so on, will help identify the structures of the bones. Sometimes it is easier to learn the bones first, then come back and learn the "bumps" and "knobs," and then go back to the pictures of the bones. Take a look at the chapter and decide what method would work best for you. Once you have the pictures down, test yourself using the multiple-choice questions, relying on descriptions and directions rather than the visual. If you first learn the visual, you can usually see it in your head so you can answer a question without the picture.

INTRODUCTION

Fill in the blanks with the correct answers.

1. The adult skeleton is composed of (how many) ____________________ separate bones.

FC 2. Bones are organized into two major subdivisions: the ____________________ and the ____________________.

3. Eighty bones make up the ____________________ and 126 bones make up the ____________________.

FC *Matching – match each description with the correct term.*

	Term		Description
4. ______	Tuberosity	a.	The main portion of a bone
5. ______	Meatus	b.	Raised bump, usually for muscle attachment
6. ______	Epicondyle	c.	Rounded bump that usually fits into a fossa
7. ______	Foramen	d.	An elongated depression or groove
8. ______	Facet	e.	A raised area or projection
9. ______	Condyle	f.	Tube-like opening or channel; does not go all the way through
10. ______	Fossa	g.	Flat surface that forms a joint with another bone
11. ______	Process	h.	Round hole for vessels and nerves
12. ______	Sulcus	i.	Depression that often receives an articulating bone
13. ______	Body	j.	Bump just proximal to the end of the long bones

SKULL

STUDY TIP

FC How to identify and remember the skull foramen from the inferior view of the skull. Start at the most anterior portion of the basilar part of the occipital bone and move directly lateral on the right side. The foramen lacerum is the first foramen encountered. Continue to move in a clockwise pattern, going from foramen to foramen using the following mnemonic. You will end up at foramen magnum and then go in reverse for the other side. Look at a picture of the skull and this will make sense.

Lots	**L**acerum (la-sir-um)
Of	**O**vale (oh-val'ee)
Spiders	**S**pinosum (spi-know'sum)
Can	**C**arotid (ka-rha-tid)
Jump	**J**ugular
High	**H**ypoglossal

Fill in the blanks with the correct answers.

14. The two major divisions of the skull are the ____________________ and the ____________________ bones.

15. The ____________________ is a U-shaped bone below the mandible that allows for many of the unique sound productions of human speech.

16. ____________________ provide space between the cranial bones and allow for compression during birth.

17. The ____________________ bone is inferior to the perpendicular plate of the ethmoid and is part of the septum.

18. The ____________________ are mucosa-lined, air-filled spaces in the cranial bones.

FC *Short Answer*

19. Identify the six cranial bones and the number of each.

20. Identify the eight facial bones and the number of each.

21. List the four paranasal sinuses.

22. List the three auditory ossicles, superficial to profundus.

23. List the three sutures of the cranium.

Multiple choice – select the best answer.

24. Which of the following is the largest paranasal sinus and is typically involved with sinus infections?
    a. Sphenoid
    b. Ethmoid
    c. Maxillary
    d. Frontal

25. *Cheek* usually refers to what bone that makes up the lateral rim of the orbit?
    a. Zygomatic
    b. Frontal
    c. Maxilla
    d. Ethmoid

26. What suture separates the occipital from the parietal bones?
    a. Squamous
    b. Coronal
    c. Lambdoidal
    d. Sagittal

27. This bone lies just posterior to each nasal bone.
    a. Ethmoid
    b. Lacrimal
    c. Maxilla
    d. Sphenoid

28. Which bone is posterior to the sphenoid on the lateral view of the skull?
    a. Temporal
    b. Frontal
    c. Zygomatic
    d. Occipital

29. The large bump just posterior and inferior the ear is the ____________.
    a. zygomatic process
    b. styloid process
    c. mastoid process
    d. external occipital protuberance

30. The sella turcica is a saddle-shaped depression in what bone?
    a. Occipital
    b. Sphenoid
    c. Maxilla
    d. Temporal

31. This foramen is at the junction of the sphenoid, temporal, and occipital bones; just lateral to the basilar portion of the occipital bone.
    a. Rotundum
    b. Spinosum
    c. Lacerum
    d. Ovale

32. What makes up the anterior portion of the hard palate?
    a. Alveolar process
    b. Palatine process
    c. Condylar process
    d. Palatine bone

33. The condylar process of the mandible is the:
    a. part that articulates with the temporal bone.
    b. main part of the bone forming the chin.
    c. large flat surface for muscle attachment.
    d. section of bone the teeth sit in.

34. If you go to the dentist to get a filling in one of your lower teeth, the dentist would inject anesthetic near what structure?
    a. Mandibular foramen
    b. Mental foramen
    c. Condylar process
    d. Infraorbital foramen

35. Both upper and lower teeth are contained in an archlike section of bone called:
    a. palatine process.
    b. periodontal process.
    c. condylar process.
    d. alveolar process.

36. The most superior part of the ethmoid bone that is located in the cranial cavity is the:
    a. cribriform plate.
    b. crista galli.
    c. ethmoidal labyrinth.
    d. sella turcica.

Labeling – label the following diagrams.

37. Anterior view of the skull

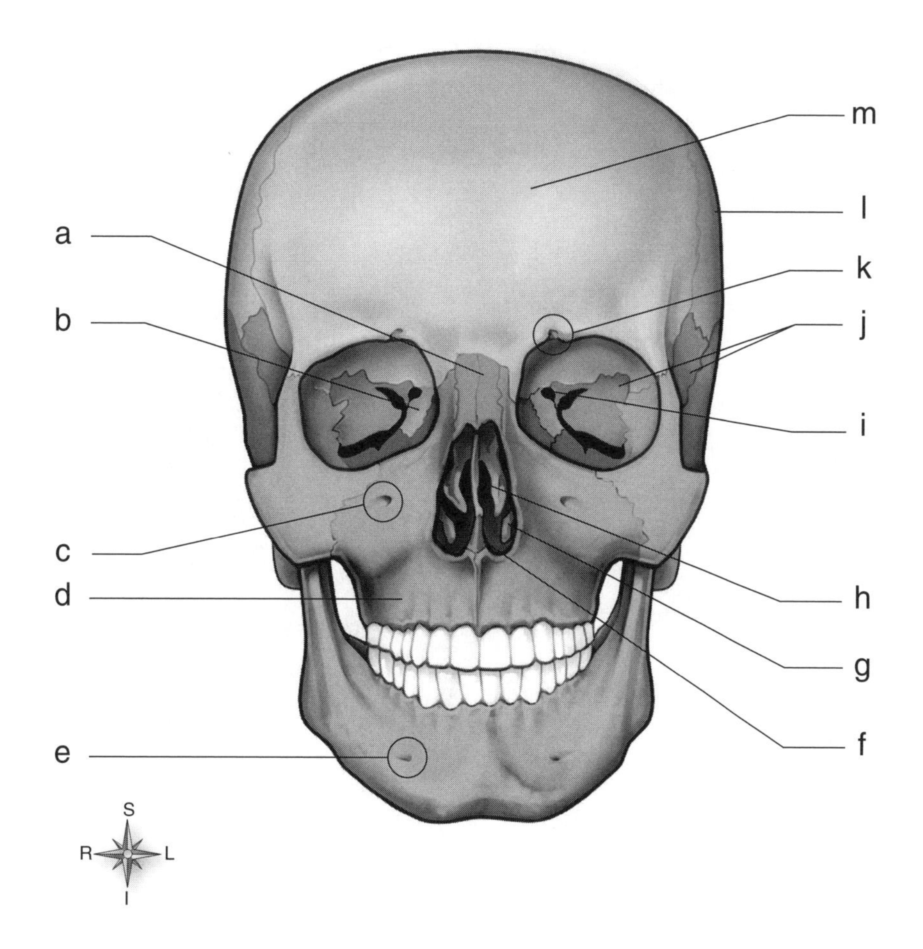

a. ______________________

b. ______________________

c. ______________________

d. ______________________

e. ______________________

f. ______________________

g. ______________________

h. ______________________

i. ______________________

j. ______________________

k. ______________________

l. ______________________

m. ______________________

38. Skull viewed from the right side

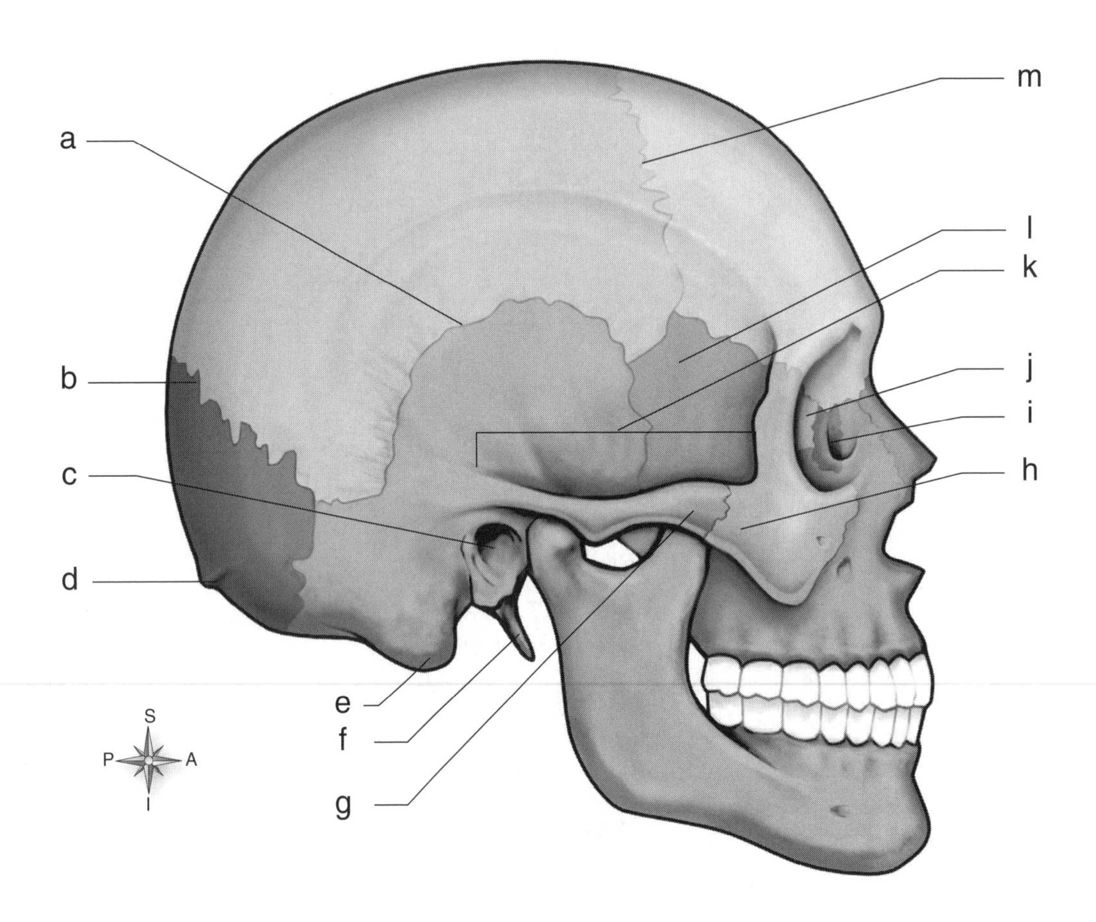

a. ______________________

b. ______________________

c. ______________________

d. ______________________

e. ______________________

f. ______________________

g. ______________________

h. ______________________

i. ______________________

j. ______________________

k. ______________________

l. ______________________

m. ______________________

39. Floor of the cranial cavity viewed from above

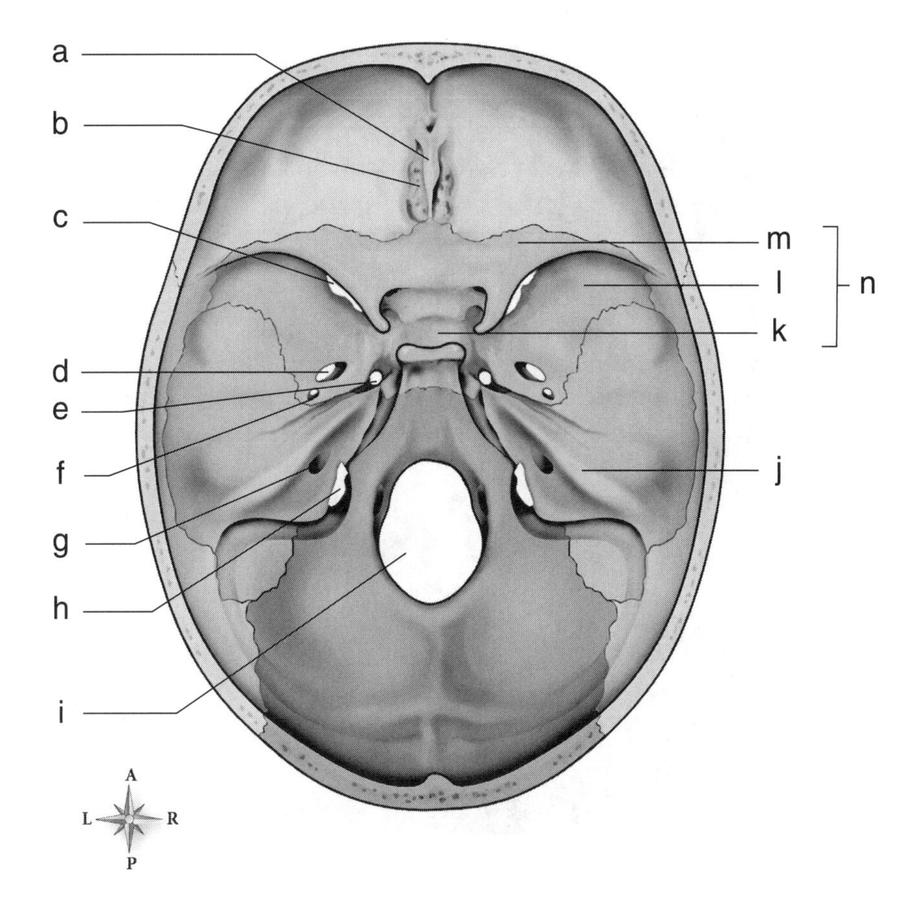

a. ______________________

b. ______________________

c. ______________________

d. ______________________

e. ______________________

f. ______________________

g. ______________________

h. ______________________

i. ______________________

j. ______________________

k. ______________________

l. ______________________

m. ______________________

n. ______________________

40. Skull viewed from below

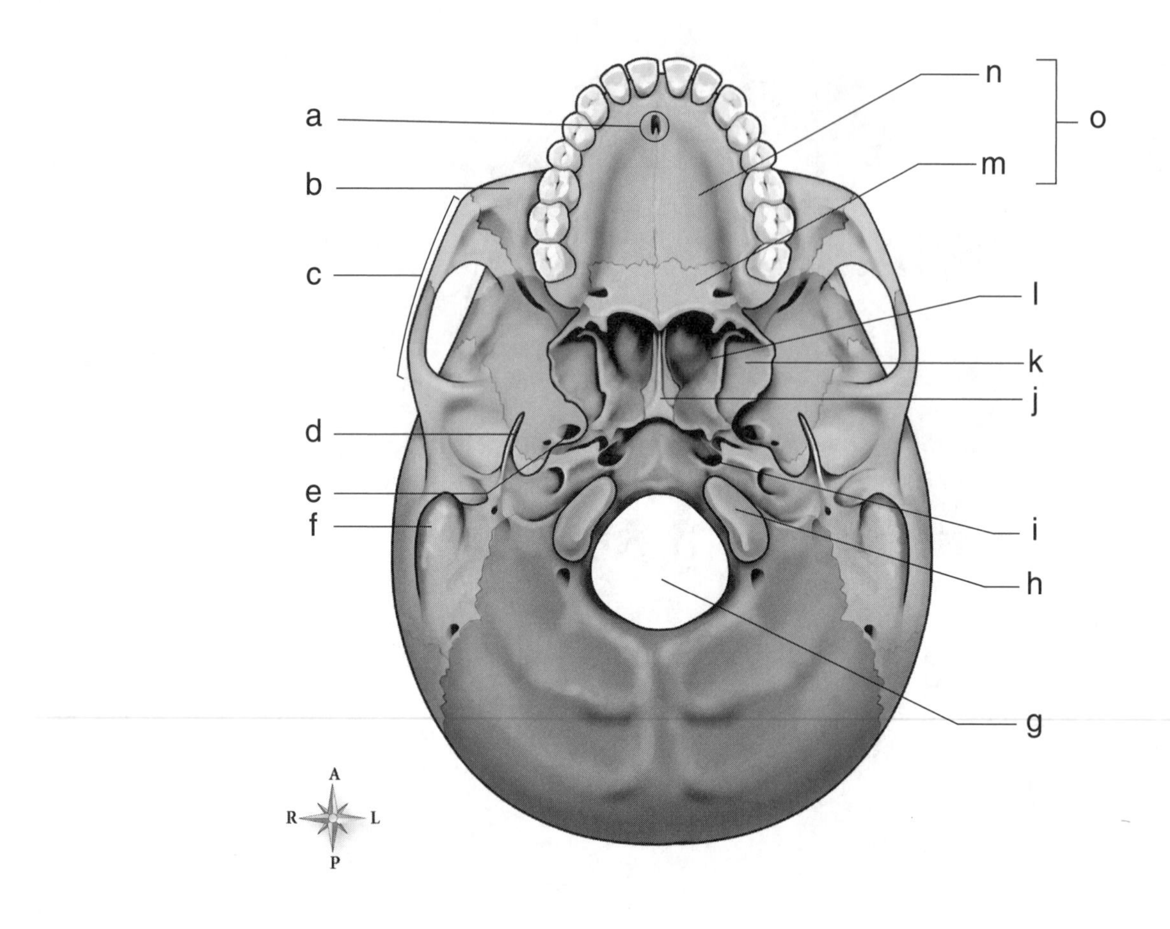

a. ____________________

b. ____________________

c. ____________________

d. ____________________

e. ____________________

f. ____________________

g. ____________________

h. ____________________

i. ____________________

j. ____________________

k. ____________________

l. ____________________

m. ____________________

n. ____________________

o. ____________________

VERTEBRAL COLUMN, STERNUM, AND RIBS

FC *Fill in the blanks with the correct answers.*

41. There are ____________________ cervical, ____________________ thoracic, and ____________________ lumbar vertebrae (numbers of each).

42. The ____________________ is a fusion of five separate vertebrae.

43. The three sections of the sternum from inferior to superior are ____________________, ____________________, and ____________________.

44. Ribs 1–7 attach directly to the costal cartilage and thus are called ____________________.

45. The ____________________ is only found on cervical vertebrae number 2 and is called the ____________________.

Labeling – label the following diagrams.

46. Cervical vertebrae (C7), superior and lateral views

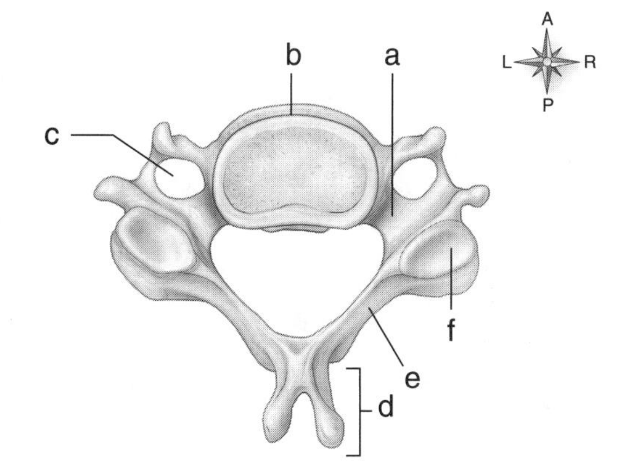

a. ______________________________

b. ______________________________

c. ______________________________

d. ______________________________

e. ______________________________

f. ______________________________

g. ______________________________

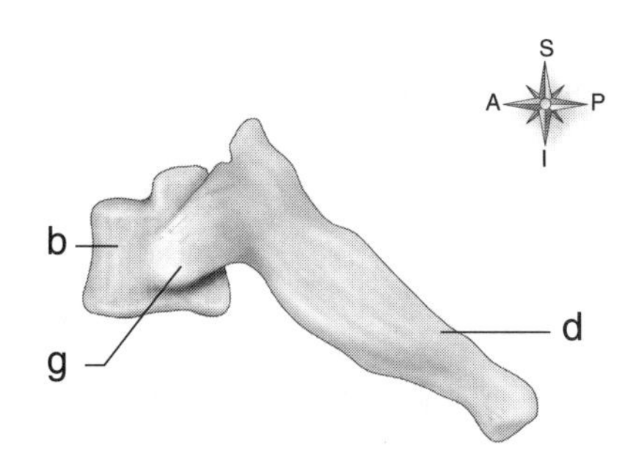

47. Thoracic vertebrae (T10), superior and lateral views

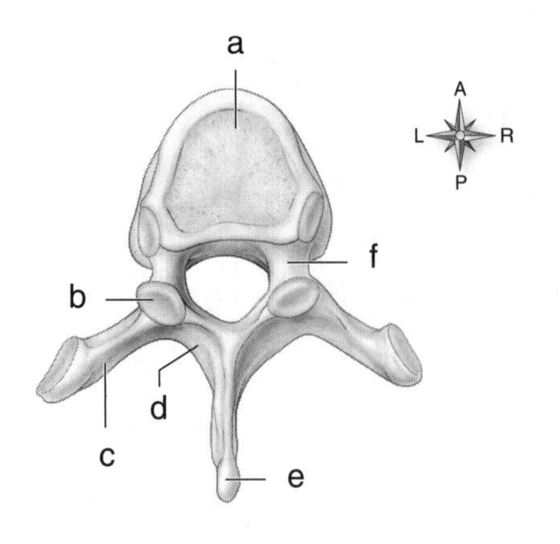

a. ____________________
b. ____________________
c. ____________________
d. ____________________
e. ____________________
f. ____________________
g. ____________________
h. ____________________

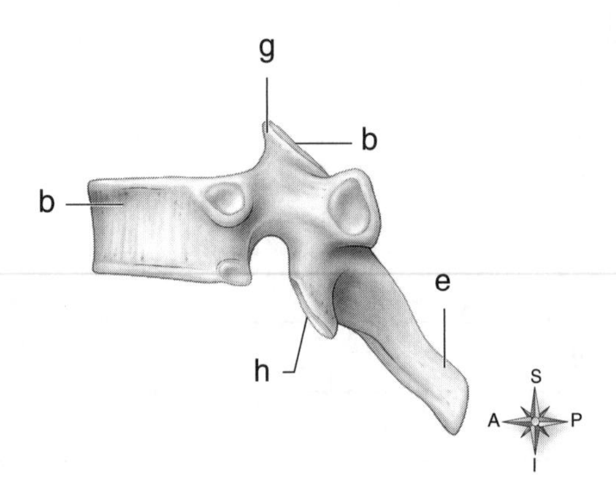

THE PECTORAL GIRDLE AND UPPER EXTREMITIES

STUDY TIP

FC Remembering the carpals—always start with the anterior view reading from the radial side medial and proximal to distal. Use the following mnemonic and give it a try.

Rowdy	**R**adial
Sam	**S**caphoid
Likes	**L**unate
To	**T**riquetrum
Push	**P**isiform
The	**T**rapezium
Toy	**T**rapezoid
Car	**C**apitate
Hard	**H**amate

You may have noticed that there are three "T's"—just remember to follow the order you just learned and "RumZiumZoid" for triquet**rum**, trape**zium**, trape**zoid**. Give it a try!

Fill in the blanks with the correct answers.

FC 48. The ____________________ consists of two clavicles and scapulae.

49. The ____________________ articulates with the radius and the ____________________ articulates with the ulna.

FC 50. There are ____________________ carpals and ____________________ metacarpals (numbers).

51. The ____________________ and ____________________ are located on the proximal end of the radius, while the ____________________ is at the distal end.

Multiple choice – select the best answer.

52. Which of the following articulates with the glenoid fossa?
    a. Spine
    b. Head of the humerus
    c. Olecranon process
    d. Greater tubercle

53. What is the name of the larger bump on the proximal end of the humerus?
    a. Greater tubercle
    b. Surgical neck
    c. Deltoid tuberosity
    d. Lesser tubercle

54. What is the most proximal structure of the ulna?
    a. Olecranon process
    b. Styloid process
    c. Head of the ulna
    d. Coronoid process

55. Which of the following is located mid-diaphysis and the lateral side of the humerus?
    a. Greater tubercle
    b. Lesser tubercle
    c. Lateral epicondyle
    d. Deltoid tuberosity

56. The ____________ is the medial condyle of the humerus.
    a. capitulum
    b. olecranon process
    c. coronoid condyle
    d. trochlea

57. Which of the following is the most anterior structure of the scapula?
    a. Coracoid process
    b. Acromion
    c. Spine
    d. Glenoid fossa

58. This is a roughened bump on the ulnar side of the radius just distal to the head.
    a. Ulnar notch of the radius
    b. Radial tuberosity
    c. Dorsal radial tuberosity
    d. Styloid process

59. Which carpal is directly medial to the scaphoid?
    a. Lunate
    b. Trapezoid
    c. Triquetrum
    d. Capitate

60. When you place your hand on the top of your shoulder, what are you most likely to feel?
    a. Glenoid fossa
    b. Spine
    c. Coracoid process
    d. Acromion

61. What is the name of the large fossa on the posterior, distal end of the humerus?
    a. Glenoid fossa
    b. Coracoid fossa
    c. Olecranon fossa
    d. Coronoid fossa

Matching – match each term with the correct description.

	Term		Definition
62. ______	Capitulum	a.	Projection at the lateral end of the scapular spine
63. ______	Deltoid tuberosity	b.	Arm socket
64. ______	Olecranon fossa	c.	Rough area about midway down the shaft on the humerus
65. ______	Radial tuberosity	d.	Rounded knob distal to the lateral epicondyle
66. ______	Phalanges	e.	Large depression on the distal posterior surface of the humerus
67. ______	Styloid process	f.	Roughened projection on the proximal anterior surface; where the biceps muscle inserts
68. ______	Olecranon process	g.	Sharp protuberance at the distal ends of the radius and ulna
69. ______	Glenoid cavity	h.	Long bones of the fingers
70. ______	Coronoid fossa	i.	"Elbow"
71. ______	Acromion	j.	Small depression on the distal anterior surface of the humerus

Labeling – label the following diagrams.

72. Right scapula, anterior view

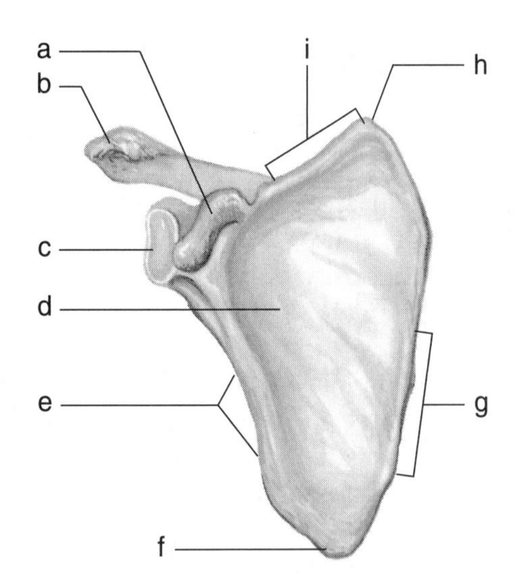

a. ______________________________

b. ______________________________

c. ______________________________

d. ______________________________

e. ______________________________

f. ______________________________

g. ______________________________

h. ______________________________

i. ______________________________

73. Right scapula, posterior view

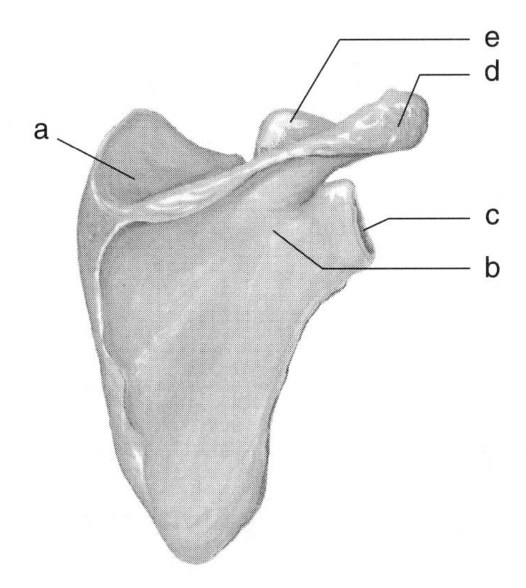

a. ______________________________

b. ______________________________

c. ______________________________

d. ______________________________

e. ______________________________

74. Right scapula, lateral view

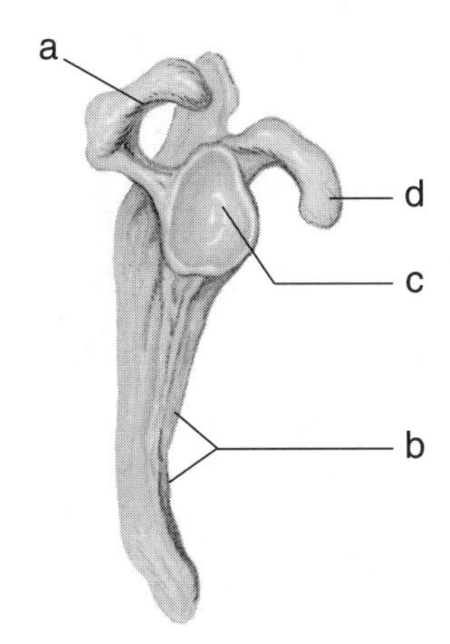

a. ______________________________

b. ______________________________

c. ______________________________

d. ______________________________

75. Humerus, anterior view

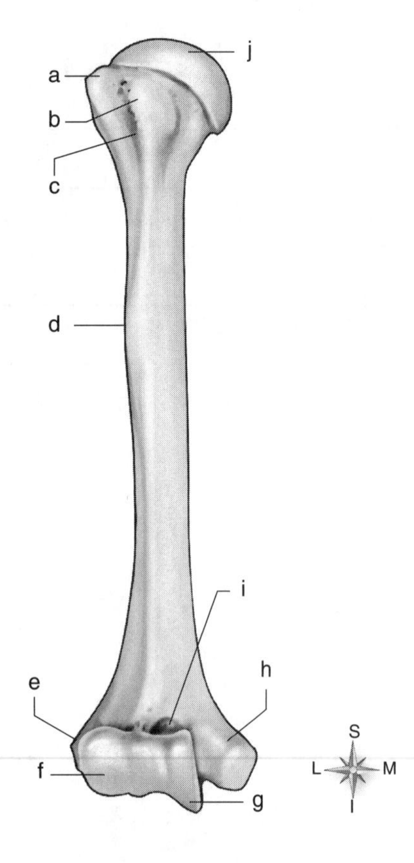

a. ______________________

b. ______________________

c. ______________________

d. ______________________

e. ______________________

f. ______________________

g. ______________________

h. ______________________

i. ______________________

j. ______________________

76. Radius and ulna, anterior view

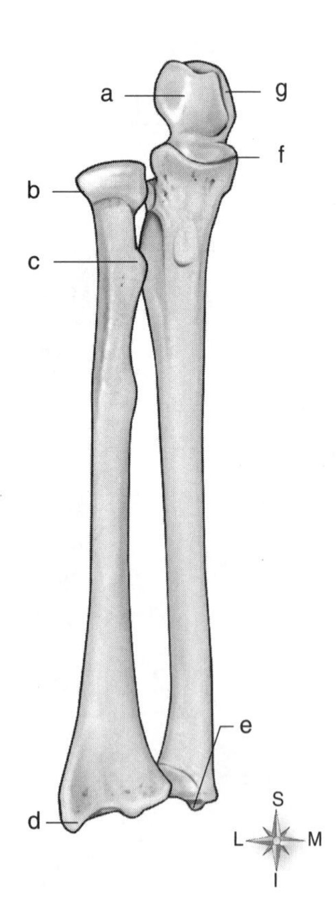

a. ______________________

b. ______________________

c. ______________________

d. ______________________

e. ______________________

f. ______________________

g. ______________________

77. Humerus, posterior view

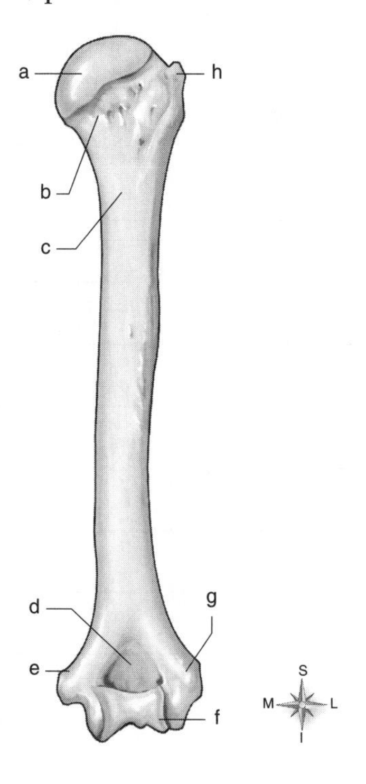

a. ______________________

b. ______________________

c. ______________________

d. ______________________

e. ______________________

f. ______________________

g. ______________________

h. ______________________

78. Radius and ulna, posterior view

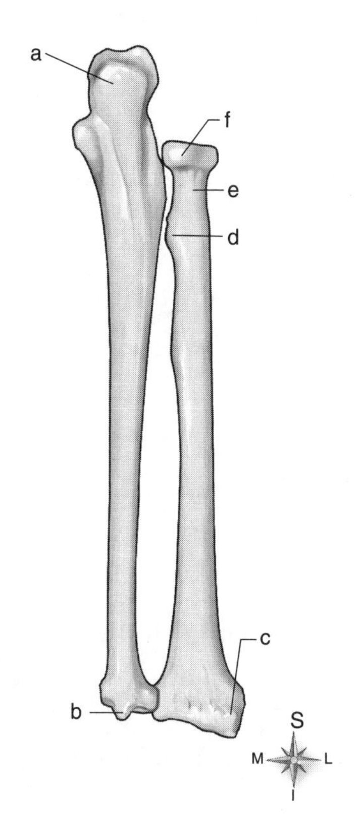

a. ______________________

b. ______________________

c. ______________________

d. ______________________

e. ______________________

f. ______________________

79. Bones of the hand and wrist

a. ______________________________

b. ______________________________

c. ______________________________

d. ______________________________

e. ______________________________

f. ______________________________

g. ______________________________

h. ______________________________

i. ______________________________

j. ______________________________

THE PELVIC GIRDLE AND LOWER EXTREMITIES

STUDY TIP

FC Remembering the tarsals—start posterior to anterior and medial to lateral and use the following mnemonic.

Can	**C**alcaneus
The	**T**alus
Nurse	**N**avicular
Medicate	**M**edial cuneiform
Insane	**I**ntermediate cuneiform
Little	**L**ateral cuneiform
Charlie	**C**uboid

Fill in the blanks with the correct answers.

80. The ____________________ is the most anterior bone of the pelvic girdle.

81. The head of the femur fits medially into the ____________________ of the coxa.

82. Of the two lower leg bones, the ____________________ is the larger and more superficial and the ____________________ is the smaller and more lateral.

83. The ____________________ articulates with the tibia and fibula.

84. The ____________________ is the largest sesamoid bone in the body.

FC *Short Answer*

85. List the three bones that make up the coxae.

86. List the three long bones that comprise the leg.

87. List the seven tarsals from posterior to anterior and medial to lateral.

Multiple choice – select the best answer.

88. The largest and most superior of the coxal bones is the:
    a. ischium.
    b. ilium.
    c. pubis.
    d. sacrum.

89. The larger projection located at the proximal end of the femur is the:
    a. greater tubercle.
    b. lesser tubercle.
    c. greater trochanter.
    d. lesser trochanter.

90. What is the most inferior coxal bone?
    a. Sacrum
    b. Ischium
    c. Pubis
    d. Ilium

91. Which of the following is directly anterior and superior to the calcaneus?
    a. Cuboid
    b. Talus
    c. Navicular
    d. Lateral cuneiform

92. What is the large "bump" on the medial side of your ankle?
    a. Medial epicondyle
    b. Medial condyle
    c. Medial malleolus
    d. Greater trochanter

93. Which of the following is located in the acetabulum?
    a. Lateral malleolus
    b. Head of the femur
    c. Obturator foramen
    d. Ischial tuberosity

94. What is the prominent projection at the anterior end of the iliac crest and can be felt externally?
    a. Ischial spine
    b. Acetabulum
    c. Anterior superior iliac spine
    d. Ischial tuberosity

95. What is the prominent ridge extending lengthwise down the posterior surface of the femur?
    a. Linea aspera
    b. Supracondylar ridge
    c. Intercondylar eminence
    d. Crest

96. The midline projection on the anterior proximal end of the tibia is the:
    a. medial condyle.
    b. tibial tuberosity.
    c. medial malleolus.
    d. intercondylar eminence.

97. What is the small projection just above the medial condyle of the femur and marks the termination of the medial supracondylar ridge?
    a. Trochlea
    b. Linea aspera
    c. Intertrochanteric line
    d. Adductor tubercle

Matching – match each term with the correct description.

	Term		Definition
98. ______	Pelvic inlet	a.	Large notch on the posterior surface of the ilium just below the posterior inferior spine
99. ______	Pubic symphysis	b.	Superior projection on the articular surface between the condyles
100. ______	Trochlea	c.	Pointed projection just above the ischial tuberosity
101. ______	Intercondylar fossa	d.	Part of the pubis lying between the symphysis and acetabulum forming the upper part of the obturator foramen
102. ______	Intercondylar eminence	e.	Opening formed by the pubic crests, iliopectineal lines, and sacral promontory; has obstetric importance
103. ______	Greater sciatic notch	f.	Smooth depression between the condyles on the anterior surface, articulating with the patella
104. ______	Crest	g.	Curve formed by the two inferior pubic rami
105. ______	Superior pubic ramus	h.	Sharp ridge on the anterior surface
106. ______	Pubic arch	i.	Deep depression between the condyles on the posterior surface
107. ______	Ischial spine	j.	Amphiarthrotic joint between pubic bones

Labeling – label the following diagrams.

108. Left coxal bone disarticulated, lateral view

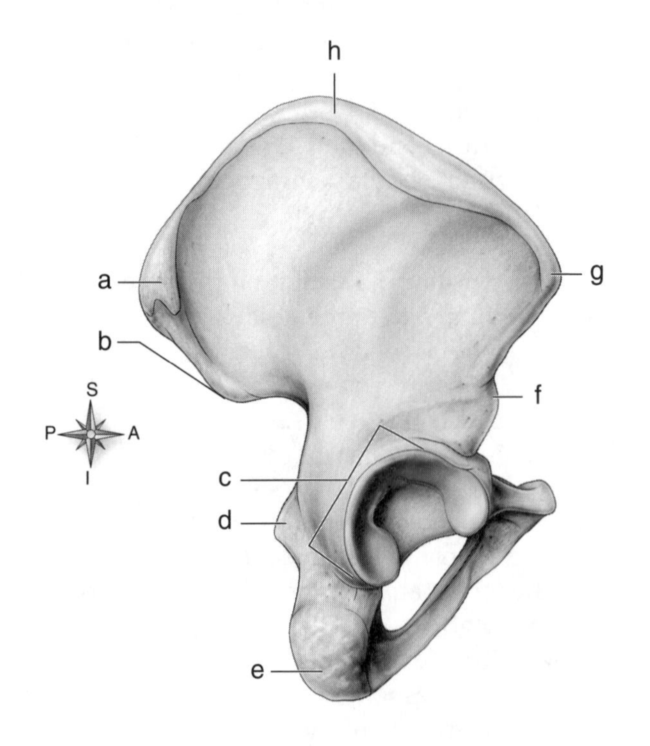

a. ______________________________

b. ______________________________

c. ______________________________

d. ______________________________

e. ______________________________

f. ______________________________

g. ______________________________

h. ______________________________

109. Right femur, anterior and posterior views

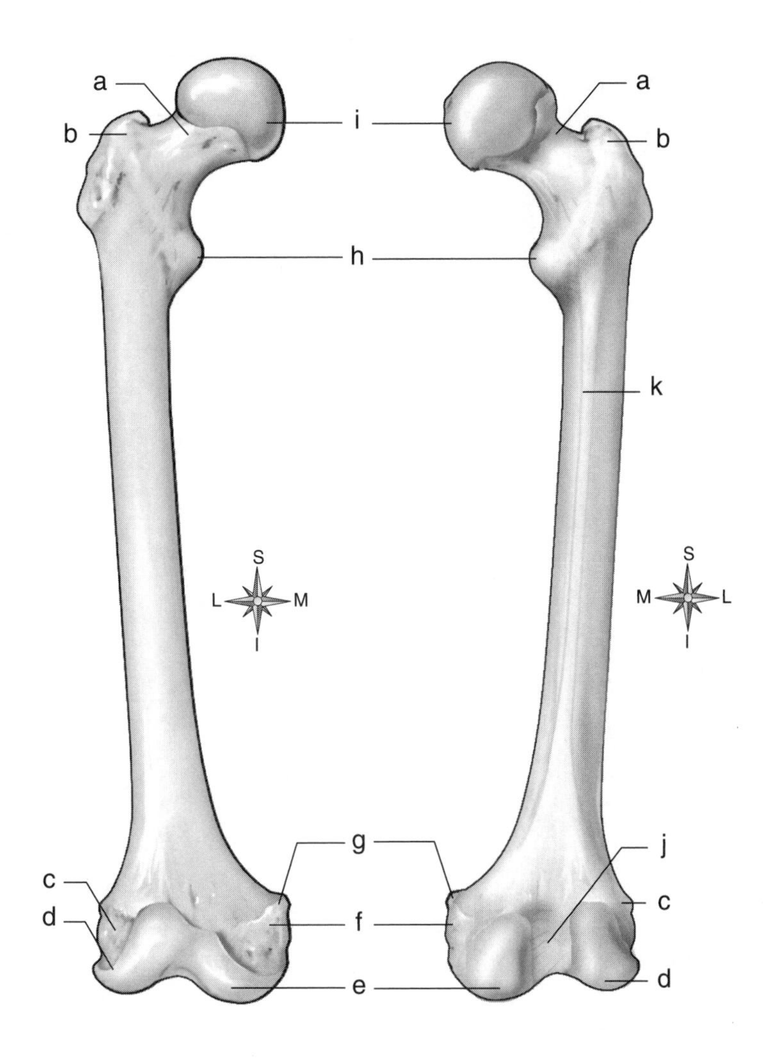

a. ______________________

b. ______________________

c. ______________________

d. ______________________

e. ______________________

f. ______________________

g. ______________________

h. ______________________

i. ______________________

j. ______________________

k. ______________________

110. Right tibia and fibula, anterior view

a. ______________________________

b. ______________________________

c. ______________________________

d. ______________________________

e. ______________________________

f. ______________________________

g. ______________________________

111. Bones of the right foot, viewed from above

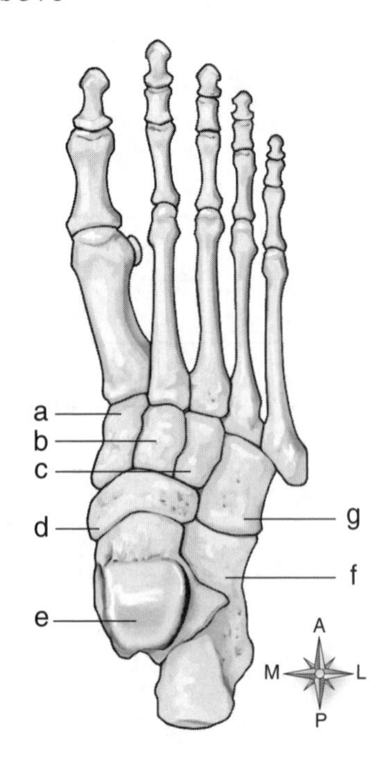

a. ______________________________

b. ______________________________

c. ______________________________

d. ______________________________

e. ______________________________

f. ______________________________

g. ______________________________

ARTICULATIONS

STUDY TIP

When learning the ligaments of the joints, simply think of what they connect to and you usually can figure out the name. Example: iliofemoral (ilium and femur), ischialfemoral (ischium and femur), acromioclavicular (acromian and clavical), coracoacromial (coracoid process and acromion). Ligaments on the sides are called *medial* and *lateral collaterals*.

Fill in the blanks with the correct answers.

112. ____________________ are joints in which fibrous bands (ligaments) connect two bones, such as between the distal ends of the radius and ulna.

FC 113. A(n) ____________________ is a point of contact between two or more bones.

FC 114. When describing joints, ____________________ are immovable; *amphiarthroses* are slightly movable, and ____________________ are freely movable.

115. Some cartilaginous joints, the ____________________, have hyaline cartilage between the articulating bones.

116. ____________________ joints are characterized by relatively flat articulating surfaces that allow limited gliding movements.

117. ____________________ joints and ____________________ joints permit movement around only one axis and in only one plane.

118. ____________________ are unique joints that occur between the root of a tooth and the mandible or maxilla.

Multiple choice – select the best answer.

FC 119. What are closely knit joints found only in the skull?
 a. Syndesmoses
 b. Sutures
 c. Gomphoses
 d. Symphyses

120. Which of the following is an example of a multiaxial joint?
 a. Elbow
 b. Shoulder
 c. Knee
 d. Thumb

121. Which of the following is a joint in which a pad or disk of fibrocartilage connects two bones?
 a. Symphysis
 b. Synchondroses
 c. Syndesmoses
 d. Synarthroses

122. Which of the following is the most mobile joint?
 a. Elbow
 b. Knee
 c. Hip
 d. Shoulder

123. Medial and lateral menisci are associated with what joint?
    a. Shoulder
    b. Elbow
    c. Knee
    d. Hip

124. Saddle and condyloid are examples of what specific type of joint?
    a. Biaxial
    b. Uniaxial
    c. Multiaxial
    d. Monoaxial

125. What instrument is used to measure range of motion (ROM)?
    a. Goniometer
    b. Angular square
    c. Motion meter
    d. Flexometer

126. A moist, slippery membrane that lines the inner surface of the joint capsule is called the:
    a. joint membrane.
    b. articular membrane.
    c. synovial membrane.
    d. bursae.

FC ***Matching – match each description with the correct term.***

	Term		Description
127. ______	Joint capsule	a.	Pads of fibrocartilage located between articulating ends of bones
128. ______	Articular cartilage	b.	Pillow-like structures that consist of a synovial membrane filled with synovial fluid
129. ______	Diarthroses	c.	A complete casing around the ends of the bones
130. ______	Ligaments	d.	Slightly movable joint
131. ______	Joint cavity	e.	A moist, slippery membrane that lines the inner surface of the joint capsule
132. ______	Amphiarthroses	f.	Immovable joint
133. ______	Synovial membrane	g.	Space between the articulating surfaces of bones
134. ______	Menisci	h.	Strong cords of dense connective tissue between bones lashing them firmly together
135. ______	Bursae	i.	Freely movable joints
136. ______	Synarthroses	j.	Thin layer of hyaline cartilage that covers and cushions the ends of bones

CHAPTER 10

Muscular System

HOW TO APPROACH THE MUSCULAR SYSTEM

This chapter is where you are really going to start using your understanding of chemistry (ions) in an applicable way. The chapter starts off with the physiology of muscles. You should be able to follow (describe in detail) a nerve impulse from the nerve, to the muscle, into the muscle, and what is happening in the muscle to cause a muscle contraction. Make sure you are solid with the basic terms such as *origin, insertion,* and *action* (see skeletal muscle structure).

A major part of this chapter is learning all the skeletal muscles. That can be much more difficult than it needs to be. Focus on the name of the muscle and where it is located. The name frequently tells you its location. Sternocleidomastoid muscle, for example, connects the sternum and clavicle to the base of the skull (mastoid process). You will be required to know the origin, insertion, action, and many times the innervation of several, if not all, the muscles. If you know its location, you really do not need to memorize origin, insertion, and action because you can figure these out just knowing its location. If I know the location but cannot remember the origin, insertion, or action, all I have to do is place a hand over the muscle and move until I contract the muscle. Now I know the action. If I know the action and the definition of *origin* and *insertion* then I simply contract the muscle and use my hand to follow the muscle and tendons to the origin(s) and insertion. This may not give you the exact locations of all origins and insertions, but it will allow you to answer most test questions. Doing this can save you vast amounts of time and effort.

INTRODUCTION TO SKELETAL MUSCLES

Fill in the blanks with the correct answers.

FC 1. Muscle ____________________ are actually individual muscle cells.

2. ____________________ allow electrical signals to move deeper into the cell and are inward extensions of the sarcolemma.

FC 3. Each ____________________ lies between two successive Z-lines.

4. When acetylcholine fuses with receptor sites on the ____________________, it opens ____________________ ion gates.

FC 5. The ____________________ is where a motor neuron comes in contact with the sarcolemma.

6. Groups of muscle fibers, called ____________________, are bound together by a thicker connective tissue called ____________________.

7. The fibrous wrapping of a muscle may extend as a broad, flat sheet of connective tissue called a(n) ____________________.

FC *Matching – match each description with the correct term.*

		Term		Definition
8.	______	Myosin heads	a.	Bind to actin molecules
9.	______	Acetylcholinesterase	b.	Between Z-lines
10.	______	Excitability	c.	Creates choline and acetate
11.	______	Muscle fiber	d.	Muscle cell membrane
12.	______	Sarcolemma	e.	Contain the sarcomeres
13.	______	Sarcoplasm	f.	Muscle cell
14.	______	Sarcoplasmic reticulum	g.	Extensions of the sarcolemma
15.	______	T-tubules	h.	Store calcium ions
16.	______	Myofibrils	i.	Muscle cytoplasm
17.	______	Sarcomere	j.	Ability to respond to nervous signals
18.	______	Perimysium	k.	Covers entire muscle
19.	______	Endomysium	l.	Binds fascicles together
20.	______	Epimysium	m.	Covers muscle fibers

Multiple choice – select the best answer.

21. Another name for a muscle fiber is a:
    a. T tubule.
    b. myofibril.
    c. myofilament.
    d. muscle cell.

22. The muscle cell membrane is called:
    a. sarcoplasmic reticulum.
    b. sarcoplasm.
    c. myomembrane.
    d. sarcolemma.

23. Sarcomeres are segments in which of the following structures?
    a. Myofilaments
    b. Muscle fibers
    c. Myofibrils
    d. T tubules

24. What is the basic contractile unit of muscle?
    a. Sarcomere
    b. Sarcoplasmic reticulum
    c. Myofilament
    d. Myofibril

25. Which of the following causes the visible "striations" in muscle cells?
    a. A-bands
    b. Z-lines
    c. T-tubules
    d. E-bands

26. The very small fibers that are classified as thick and thin filaments are called:
    a. Z-lines.
    b. myofilaments.
    c. myofibrils.
    d. myofibers.

27. Which of the following stores calcium ions in the muscle cell?
    a. Sarcoplasm
    b. Sarcolemma
    c. Sarcoplasmic reticulum
    d. Sarcomere

28. Which of the following is NOT a protein molecule that makes up the myofilament?
    a. Tropomyosin
    b. Troponin
    c. Myosin
    d. Actosin

29. Which of the following is the neurotransmitter that is involved with contraction of muscle?
    a. Acetylcholine
    b. Dopamine
    c. Serotonin
    d. Acetylcholinesterase

30. Calcium ions combine with what in the thin filaments exposing the active binding sites?
    a. Actin
    b. Myosin
    c. Troponin
    d. Tropomyosin

31. Once the active sites for contraction are exposed, what binds to actin molecules?
    a. Myosin heads
    b. Actin heads
    c. Troponin heads
    d. Tropomyosin heads

32. What is responsible for breaking down the neurotransmitter in the neuromuscular junction (NMJ) into an acetate group and choline?
    a. Acetylcholine
    b. ANP
    c. Acetylcholinesterase
    d. ATPase

33. Muscle fibers are covered by which of the following?
    a. Fascicles
    b. Fascia
    c. Endomysium
    d. Epimysium

34. Epimysium, perimysium, and endomysium may join with fibrous tissue that extends from the muscle, forming:
    a. fascia.
    b. fascicles.
    c. a tendon.
    d. a ligament.

35. The model of muscle contraction is called the:
    a. sliding filament model.
    b. contractile filament model.
    c. ratcheting model.
    d. myosin head model.

Labeling – label the following diagrams.

36. Structures of a skeletal muscle

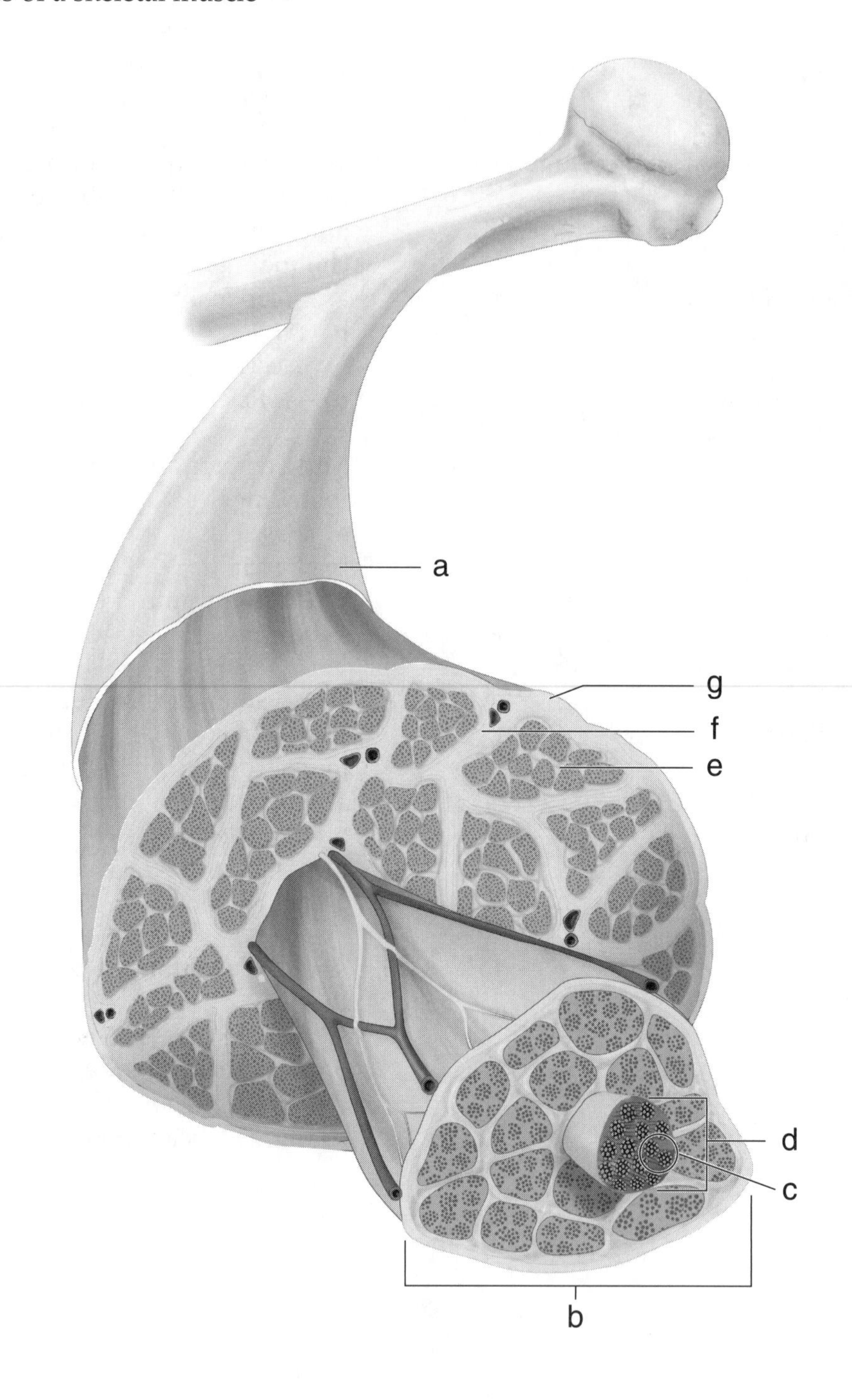

a. ______________________

b. ______________________

c. ______________________

d. ______________________

e. ______________________

f. ______________________

g. ______________________

37. Unique features of the skeletal muscle cell

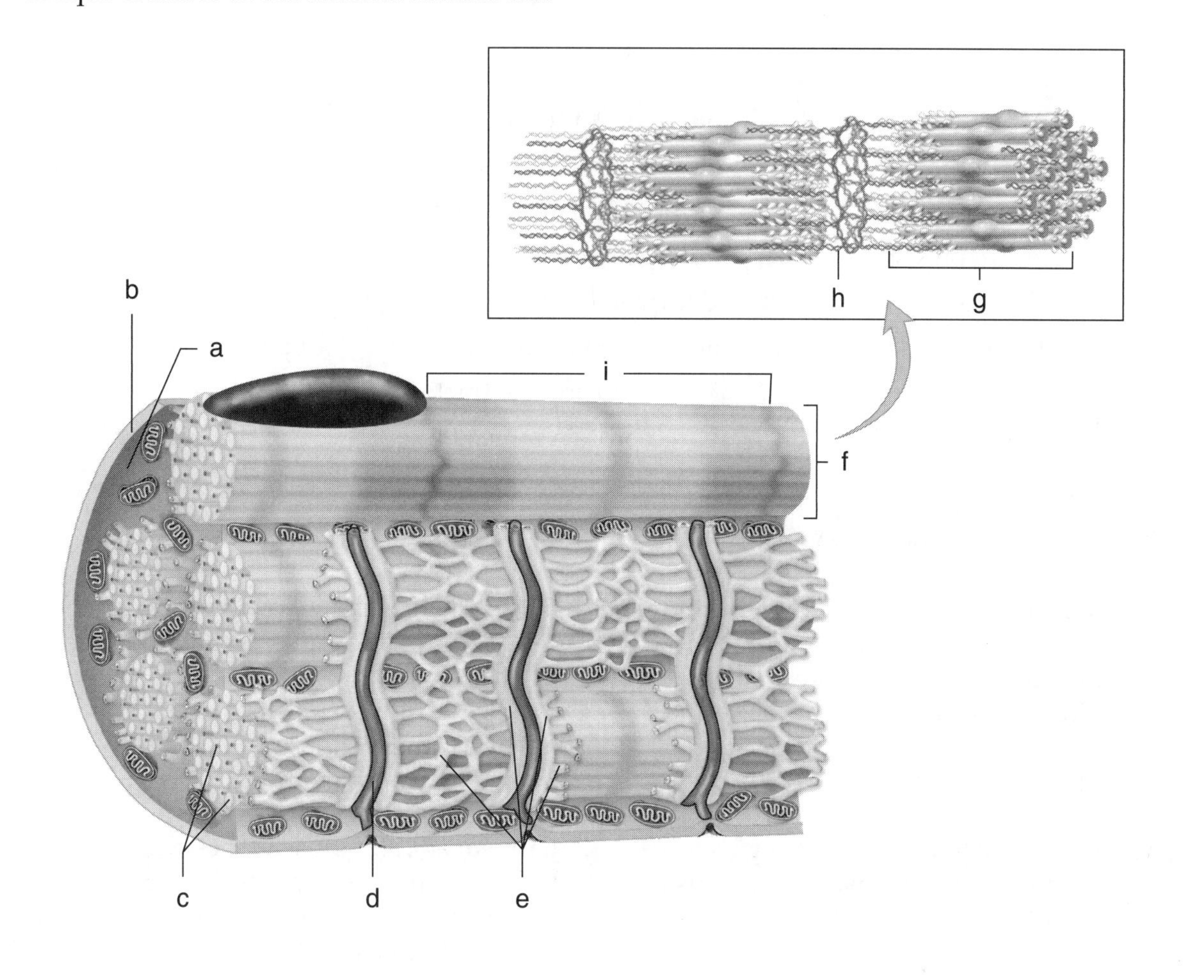

a. ______________________

b. ______________________

c. ______________________

d. ______________________

e. ______________________

f. ______________________

g. ______________________

h. ______________________

i. ______________________

SKELETAL MUSCLES: ENERGY SOURCES, FUNCTION, AND STRUCTURES

Fill in the blanks with the correct answers.

38. Excess oxygen molecules in the sarcoplasm are bound to a large protein molecule called ______________________.

39. The ______________________ the number of muscle fibers supplied by a motor unit, the more ______________________ coordinated the movements that muscle can produce.

40. In a(n) ______________________ contraction, the muscle shortens; in a(n) ______________________ contraction, the muscle pulls against a load but does NOT shorten.

FC ***Matching – match each description with the correct term.***

	Term		Definition
41. ______	Myoglobin	a.	Surrounds muscle and tendon
42. ______	Fixator	b.	Does not move during contraction
43. ______	Synergists	c.	Opposes the movement
44. ______	Antagonists	d.	Performs the movement
45. ______	Agonist	e.	Complement the movement
46. ______	Origin	f.	Stabilize joints and help maintain posture
47. ______	Insertion	g.	Moves during contraction
48. ______	Fascia	h.	Large protein molecule

Multiple choice – select the best answer.

49. "Red meat" or "dark meat" would contain large quantities of what?
    a. Blood
    b. Blood vessels
    c. Myofibrils
    d. Myoglobin

FC 50. By going to gym and lifting weights, muscles can increase in size. This is called:
    a. atrophy.
    b. myogenesis.
    c. hypertrophy.
    d. catabolism.

51. A motor unit is a:
    a. single muscle fiber.
    b. motor neuron and the muscle fiber.
    c. single motor nerve.
    d. group of muscles with the same action.

52. Which of the following is a continual, partial contraction of a muscle?
    a. Muscle tone
    b. Tetany
    c. Hypertrophy
    d. Atrophy

FC 53. Prolonged disuse of muscles causes:
    a. hypertrophy.
    b. decrease in muscle fibers.
    c. decrease in number of red fibers.
    d. atrophy.

54. Which of the following is the point of attachment for muscle that does not move?
    a. Insertion
    b. Origin
    c. Innervations
    d. Point of fixation

55. An agonist muscle does what?
    a. Directly opposes the action
    b. Complements the actions of the antagonist
    c. Serves to stabilize joints
    d. Directly performs a specific movement

56. Which of the following best describes insertion?
    a. Moves
    b. Does not move
    c. Contracts
    d. Stabilizes

57. The central portion of a muscle is called the:
    a. fulcrum.
    b. belly.
    c. head.
    d. origin.

Short answer

58. Name the two points of muscle attachment.

59. List four ways to classify muscles according to their actions.

MUSCLES OF THE HEAD AND TRUNK

STUDY TIP

As you learn the muscles in class and at home, it is very helpful to flex or contract them so you know exactly where they are and what they do. This will help you to learn them!

Matching – match each action with the correct muscle (action may be used more than once).

	Muscle		Action
60. ______	Occipitofrontalis (frontal belly)	a.	Grinds teeth
61. ______	Occipitofrontalis (occipital belly)	b.	Moves cheeks medially
62. ______	Temporalis	c.	Closes jaw, "biting muscle"
63. ______	Orbicularis oculi	d.	Raises eyebrows and wrinkles forehead, "surprise" look
64. ______	Zygomaticus major	e.	Draws lips together, "kissing muscle"
65. ______	Orbicularis oris	f.	Draws scalp backward
66. ______	Buccinators	g.	Squints eyes
67. ______	Masseter	h.	"Smiling" muscle
68. ______	Corrugator supercilii	i.	"Frowning muscle," brings eyebrows together
69. ______	Pterygoids		

	Muscle		Action
70. ______	Transverse abdominis	a.	Flexes the head, "prayer muscle," and rotates head toward opposite side
71. ______	Semispinalis capitis	b.	Compresses abdomen
72. ______	Sternocleidomastoid	c.	Compresses abdomen and flexes trunk
73. ______	Rectus abdominis	d.	Extends the back, neck, and head
74. ______	External oblique	e.	Extends head, like looking up
75. ______	Splenius capitis		
76. ______	Internal oblique		
77. ______	Erector spinae group		

	Muscle		Action
78. _____	Rhomboid minor	a.	Pulls scapula down and forward, abducts and rotates it upward
79. _____	Trapezius	b.	Pulls shoulders down and forward
80. _____	Pectoralis minor	c.	Retracts, rotates, and elevates the scapula
81. _____	Serratus anterior	d.	Rotates the arm medially (inward)
82. _____	Rhomboid major	e.	Flexes and adducts the upper arm, "like a push-up"
83. _____	Levator scapulae	f.	Abducts the upper arm, "flu shot muscle"
84. _____	Pectoralis major	g.	Assists in abducting the arm
85. _____	Deltoid	h.	Shrugs and raises and lowers shoulders
86. _____	Latissimus dorsi	i.	Rotates the arm outward
87. _____	Supraspinatus	j.	Elevates and retracts the scapula and abducts the neck
88. _____	Subscapularis	k.	Extends the upper arm and adducts posteriorly, "swimming muscle"
89. _____	Infraspinatus	l.	Assists in extension, adduction, and medial rotation of arm
90. _____	Teres major	m.	Rotates the arm laterally (outward)
91. _____	Teres minor		

Multiple choice – select the best answer.

92. Which of the following muscles is medial and deep to the masseter?
    a. Zygomaticus major
    b. Zygomaticus minor
    c. Buccinator
    d. Corrugator supercilii

93. This muscle encircles the mouth and is sometimes called the "kissing muscle."
    a. Orbicularis oris
    b. Orbicularis oculi
    c. Pterygoid
    d. Buccinator

94. Which of the following elevates the corners of the mouth and is called the "smiling muscle"?
    a. Orbicularis oris
    b. Zygomaticus major
    c. Buccinator
    d. Pterygoid

95. Your "jaw muscle" is which of the following?
    a. Zygomaticus
    b. Corrugator supercilii
    c. Masseter
    d. Sternocleidomastoid

96. Which of the following muscles is deep to splenius capitis?
    a. Sternocleidomastoid
    b. Internal oblique
    c. Semispinalis capitis
    d. Trapezius

97. This group of muscles is what most people refer to when they say "I strained my back."
    a. Rhomboids
    b. Obliques
    c. Erector spinae
    d. Trapezius

98. Most of the time when your "shoulders hurt," what muscle are you referring to?
    a. Trapezius
    b. Semispinalis capitis
    c. Splenius capitis
    d. Latissimus dorsi

99. Many people refer to this muscle as the "chest muscle."
    a. Latissimus dorsi
    b. Pectoralis minor
    c. Pectoralis major
    d. Serratus anterior

100. This muscle originates on C6-C7 and inserts on the medial border of the scapula.
    a. Serratus anterior
    b. Levator scapulae
    c. Rhomboid minor
    d. Trapezius

101. Which of the following is NOT part of the rotator cuff?
    a. Teres major
    b. Supraspinatus
    c. Infraspinatus
    d. Subscapularis

102. Which of the following is the most lateral muscle?
    a. Trapezius
    b. Deltoid
    c. Latissimus dorsi
    d. Pectoralis major

103. This muscle gets really big in swimmers and adducts the upper arm posteriorly.
    a. Latissimus dorsi
    b. Pectoralis major
    c. Deltoid
    d. Trapezius

104. Which of the following originates on the transverse processes of cervical vertebrae and inserts on the scapula?
    a. Depressor scapulae
    b. Trapezius
    c. Serratus anterior
    d. Levator scapulae

Labeling – label the following diagrams.

105. Muscles of the head and neck

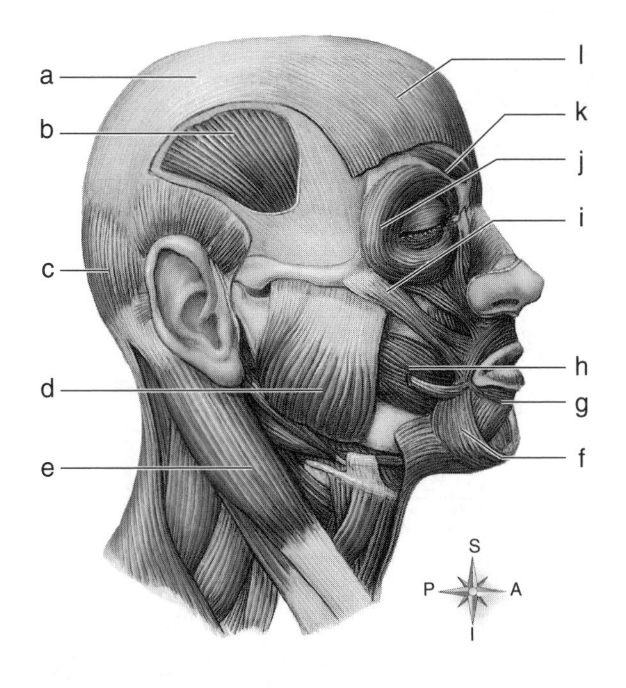

a. ______________________

b. ______________________

c. ______________________

d. ______________________

e. ______________________

f. ______________________

g. ______________________

h. ______________________

i. ______________________

j. ______________________

k. ______________________

l. ______________________

106. Muscles of the back

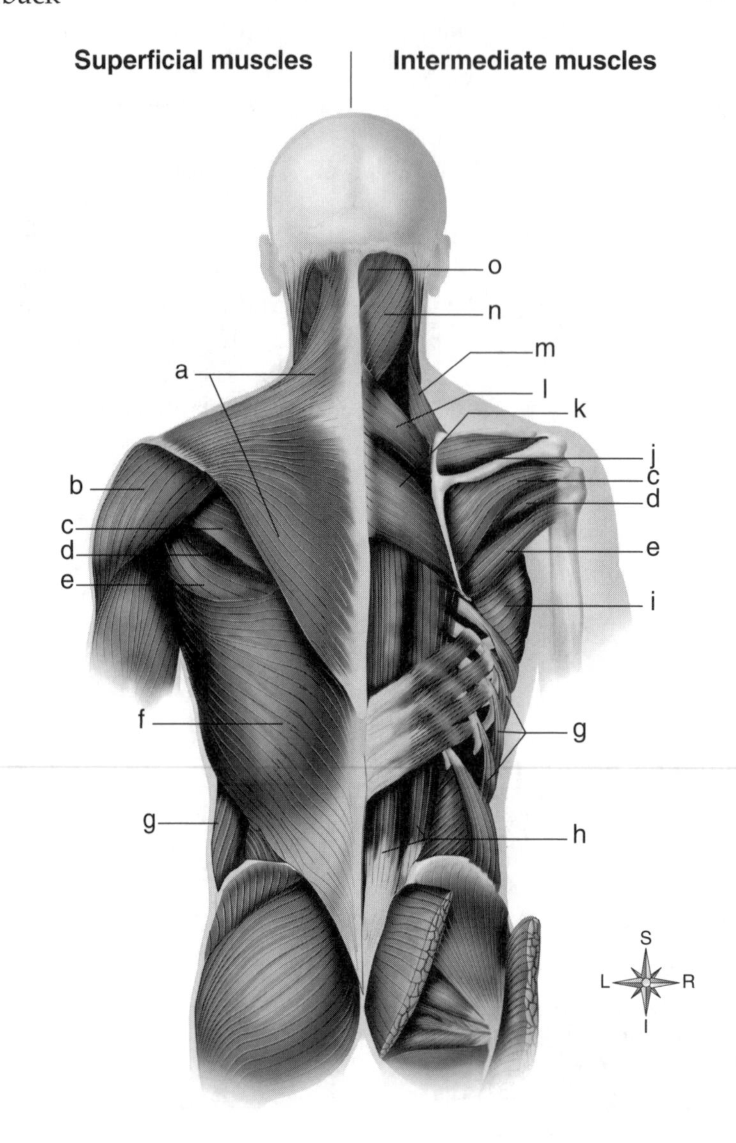

a. ______________________

b. ______________________

c. ______________________

d. ______________________

e. ______________________

f. ______________________

g. ______________________

h. ______________________

i. ______________________

j. ______________________

k. ______________________

l. ______________________

m. ______________________

n. ______________________

o. ______________________

107. Muscles of the trunk and abdominal wall

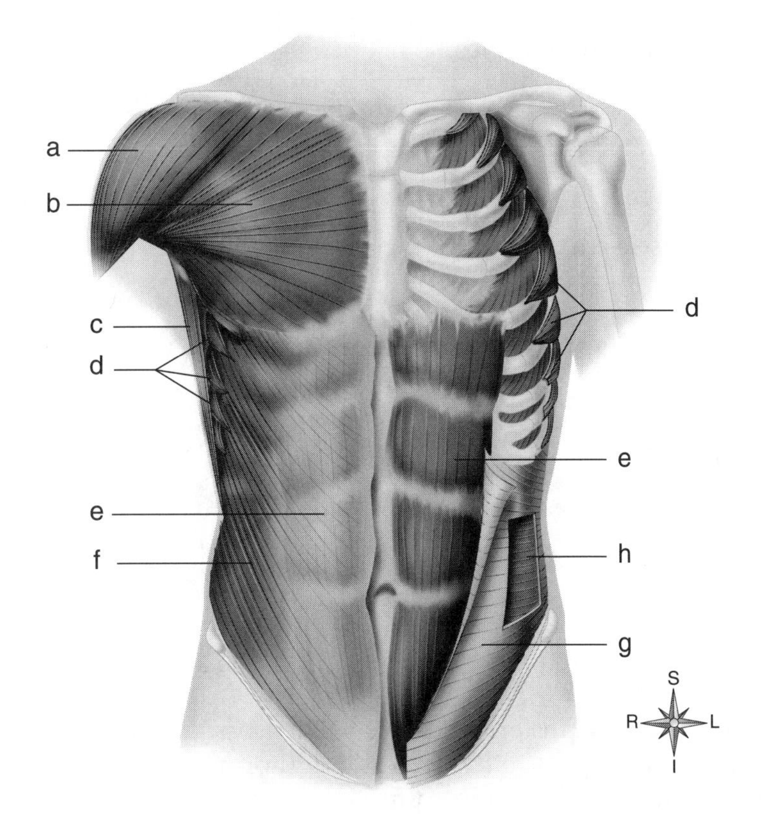

a. ______________________

b. ______________________

c. ______________________

d. ______________________

e. ______________________

f. ______________________

g. ______________________

h. ______________________

108. Rotator cuff muscles

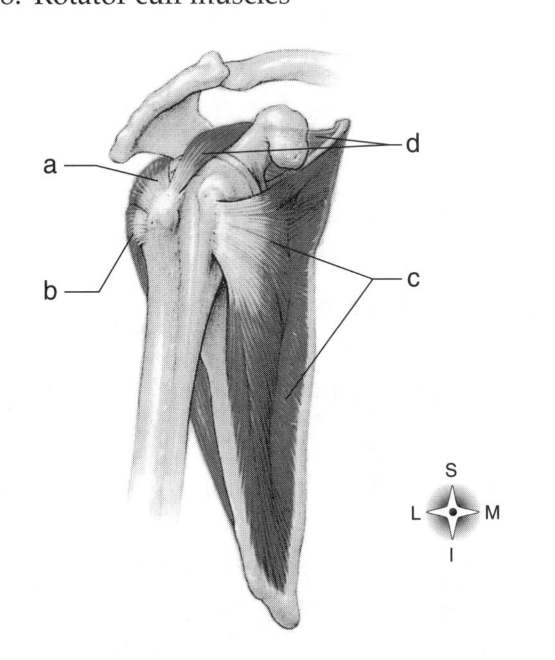

a. ______________________

b. ______________________

c. ______________________

d. ______________________

Short answer

FC 109. Name the four rotator cuff muscles.

FC 110. Name two muscles of mastication.

FC 111. Name the four muscles that make up the abdominal wall.

112. Name the two major muscles that move the upper arm.

UPPER LIMB MUSCLES

STUDY TIP

Three very important rules to help remember the muscles of the antebrachium:

1. Anterior surface = flexors, posterior surface = extensors
2. Goes to the wrist = carpi, goes to the fingers = digitorum
3. Thumb side = radialis, "pinkie" side = ulnaris

Try it!

Another thing to remember that will help you with the muscles of the appendages is the muscle that moves a part of the appendage is always proximal to what is being moved. If you want to move your humerus, the muscles are located on the shoulder. If you want to move your radius and ulna, the muscles are located on the humerus. If you want to move your hand and fingers, the muscles are located on the radius and ulna. Remembering this can help you answer some questions when you are unsure of the answer. Use this information to rule out other answers.

Matching – match each action with the correct muscle.

	Muscle		Action
113. ______	Biceps brachii	a.	Extends the fingers
114. ______	Triceps brachii	b.	Extends and abducts the hand
115. ______	Brachialis	c.	Flexes the fingers
116. ______	Flexor carpi radialis	d.	Flexes the pronated forearm
117. ______	Brachioradialis	e.	Extends the forearm
118. ______	Palmaris longus	f.	Flexes supinated forearm
119. ______	Extensor carpi radialis longus	g.	Flexes the hand only
120. ______	Extensor carpi ulnaris	h.	Flexes the hand and assists with forearm flexion
121. ______	Flexor digitorum superficialis	i.	Flexes the semipronated forearm
122. ______	Extensor digitorum	j.	Extends and adducts the hand

Multiple choice – select the best answer.

123. Which of the following muscles on the anterior surface of the antebrachium is directly medial to (next to) the brachioradialis?
    a. Flexor carpi radialis
    b. Extensor carpi radialis longus
    c. Flexor carpi ulnaris
    d. Extensor carpi ulnaris

124. What muscle is profundus to the palmaris longus?
    a. Brachioradialis
    b. Flexor carpi ulnaris
    c. Flexor digitorum superficialis
    d. Flexor carpi radialis

125. Which of the following is superficial to the brachialis?
    a. Extensor digitorum
    b. Extensor carpi ulnaris
    c. Flexor carpi ulnaris
    d. Biceps brachii

126. Which of the following is the most lateral muscle of the antebrachium?
    a. Flexor carpi radialis
    b. Brachioradialis
    c. Extensor carpi radialis longus
    d. Extensor carpi radialis brevis

127. Which of the following is medial to the palmaris longus?
    a. Flexor carpi ulnaris
    b. Flexor carpi radialis
    c. Brachioradialis
    d. Flexor digitorum superficialis

128. Which of the following is lateral to extensor carpi radialis longus?
    a. Extensor carpi radialis brevis
    b. Brachioradialis
    c. Extensor digitorum
    d. Flexor carpi radialis

129. Which of the following is medial to extensor carpi ulnaris?
    a. Extensor carpi radialis brevis
    b. Flexor carpi radialis
    c. Extensor digitorum
    d. Flexor carpi ulnaris

130. Extensor carpi radialis brevis is directly lateral to which of the following muscles?
    a. Extensor carpi radialis longus
    b. Brachioradialis
    c. Extensor digitorum
    d. Flexor carpi radialis

Labeling – label the following diagrams.

131. Muscles acting on the forearm

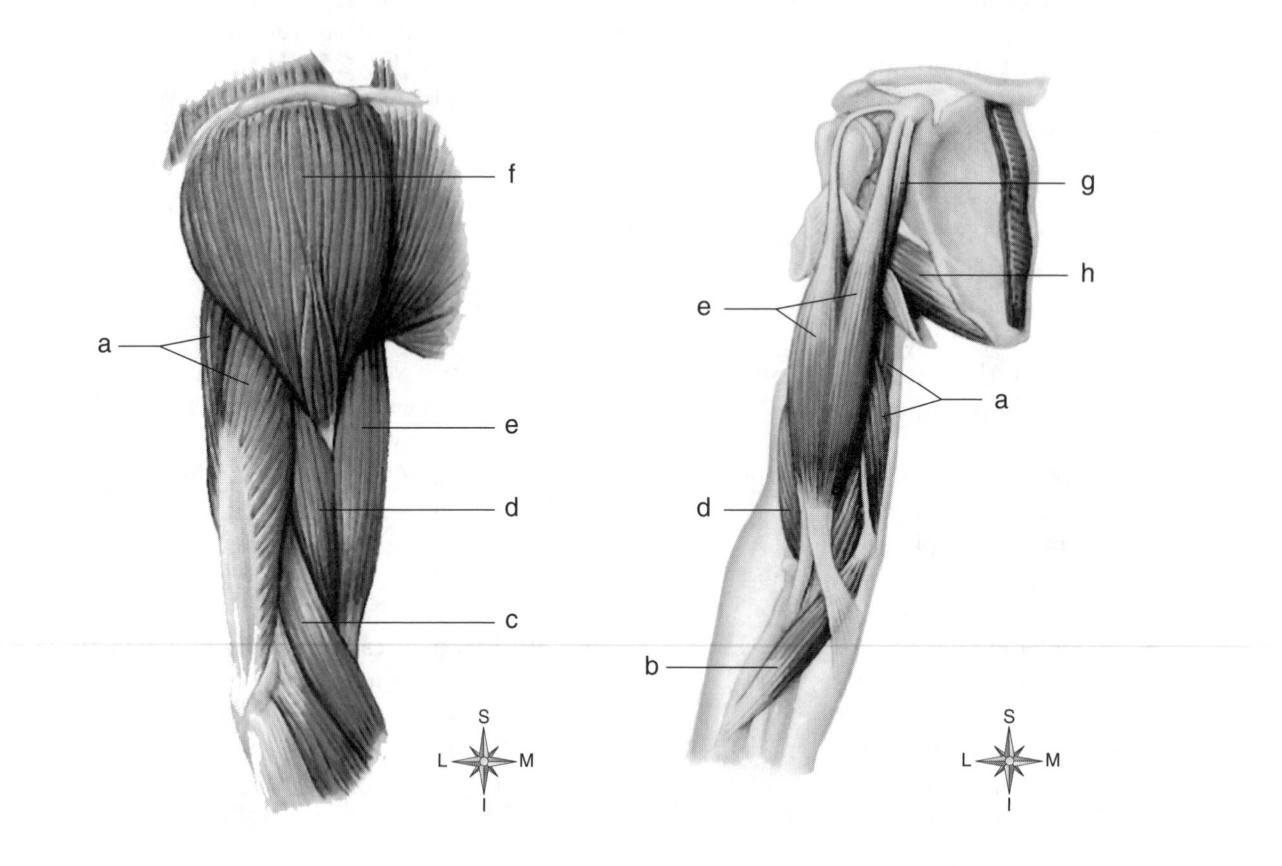

a. ______________________

b. ______________________

c. ______________________

d. ______________________

e. ______________________

f. ______________________

g. ______________________

h. ______________________

LOWER LIMB MUSCLES

Matching – match each action with the correct muscle (muscles will be used more than once).

	Muscle		Action
132. ______	Gluteus maximus	a.	Extends the leg, like kicking
133. ______	Rectus femoris	b.	Extends the thigh and flexes the leg
134. ______	Gluteus medius	c.	Extends the thigh
135. ______	Vastus lateralis	d.	Moves the foot towards the head
136. ______	Vastus medialis	e.	Abducts the thigh and rotates outward
137. ______	Biceps femoris	f.	Plantarflexes the foot
138. ______	Semitendinosus	g.	Flexes the thigh and extends lower leg
139. ______	Tibialis anterior	h.	Flexes the thigh and used in "sit-ups"
140. ______	Vastus intermedius		
141. ______	Semimembranosus		
142. ______	Gastrocnemius		
143. ______	Soleus		
144. ______	Iliopsoas		

Multiple choice – select the best answer.

145. Which of the following muscles is NOT located on the thigh?
    a. Tensor fasciae latae
    b. Gracilis
    c. Adductor magnus
    d. Sartorius

146. This muscle is usually referred to as the "butt" muscle.
    a. Gluteus medius
    b. Iliopsoas
    c. Gracilis
    d. Gluteus maximus

147. This muscle is involved in "sit-ups" and flexes the trunk when the femur acts as the origin.
    a. Rectus abdominis
    b. Rectus femoris
    c. Iliopsoas
    d. External oblique

148. What muscle lies between the vastus lateralis and the vastus medialis?
    a. Rectus femoris
    b. Biceps femoris
    c. Sartorius
    d. Gracilis

149. This muscle is a frequent site for injections and abducts the thigh.
    a. Gluteus maximus
    b. Gracilis
    c. Iliopsoas
    d. Gluteus medius

150. Infants get injections in this muscle located on the outside of the thigh.
    a. Rectus femoris
    b. Biceps femoris
    c. Vastus lateralis
    d. Vastus medialis

151. Which of the following is the most medial muscle of the hamstring group?
    a. Biceps femoris
    b. Semitendinosus
    c. Semimembranosus
    d. Vastus medialis

152. Which of the following muscles goes from superior/lateral to inferior/medial?
    a. Gracilis
    b. Tensor fasciae latae
    c. Sartorius
    d. Semimembranosus

153. Which of the following is the most medial muscle of the thigh?
    a. Gracilis
    b. Tensor fasciae latae
    c. Sartorius
    d. Semimembranosus

154. Which muscle lies directly lateral to the proximal half of the tibia?
    a. Peroneus longus
    b. Peroneus brevis
    c. Extensor digitorum longus
    d. Tibialis anterior

155. Which of the following is the most dorsal muscle?
    a. Soleus
    b. Gastrocnemius
    c. Extensor digitorum longus
    d. Flexor digitorum brevis

156. Which of the following lies directly inferior to the peroneus longus?
    a. Peroneus tertius
    b. Peroneus brevis
    c. Soleus
    d. Abductor hallucis

157. Which muscle lies directly anterior to the gastrocnemius?
    a. Tibialis anterior
    b. Soleus
    c. Peroneus longus
    d. Tibialis posterior

Labeling – label the following diagrams.

158. Muscles of the anterior aspect of the thigh

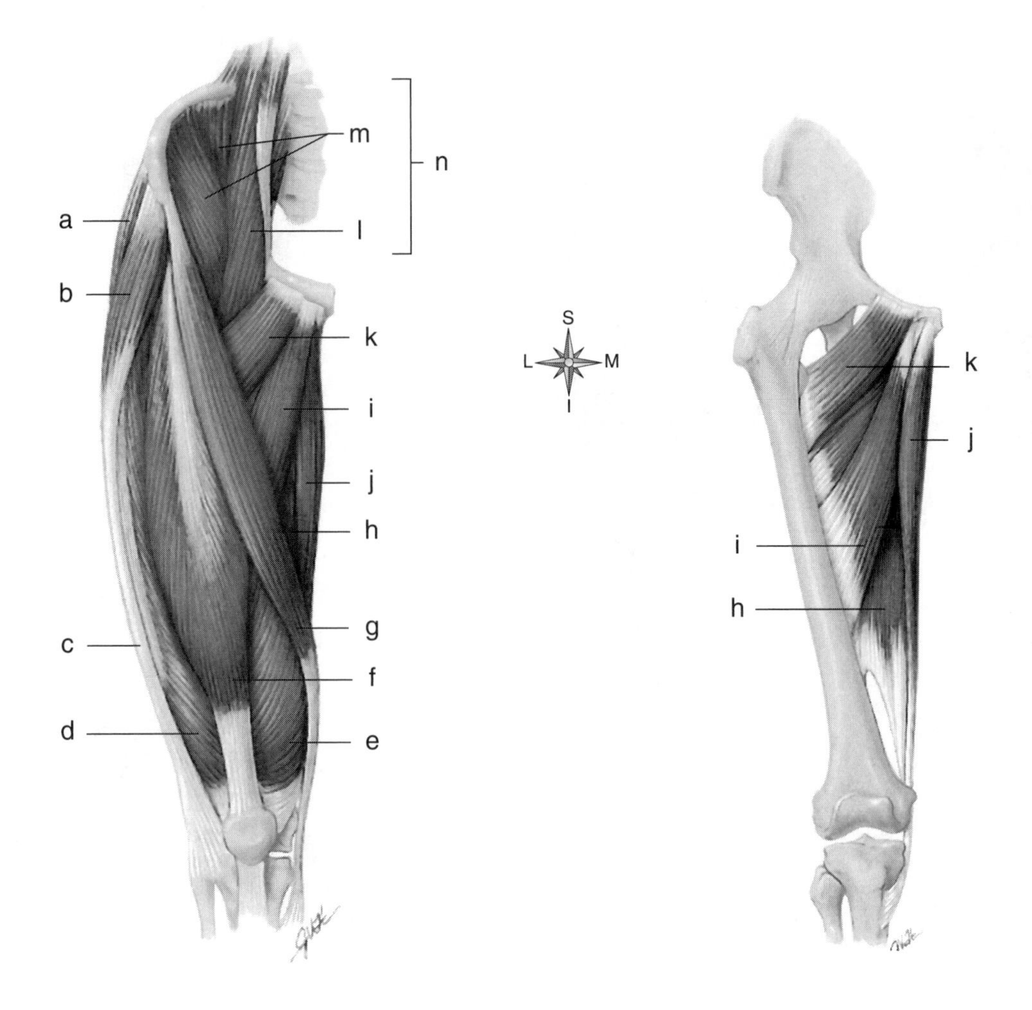

a. ______________________ h. ______________________

b. ______________________ i. ______________________

c. ______________________ j. ______________________

d. ______________________ k. ______________________

e. ______________________ l. ______________________

f. ______________________ m. ______________________

g. ______________________ n. ______________________

159. Superficial muscles of the leg

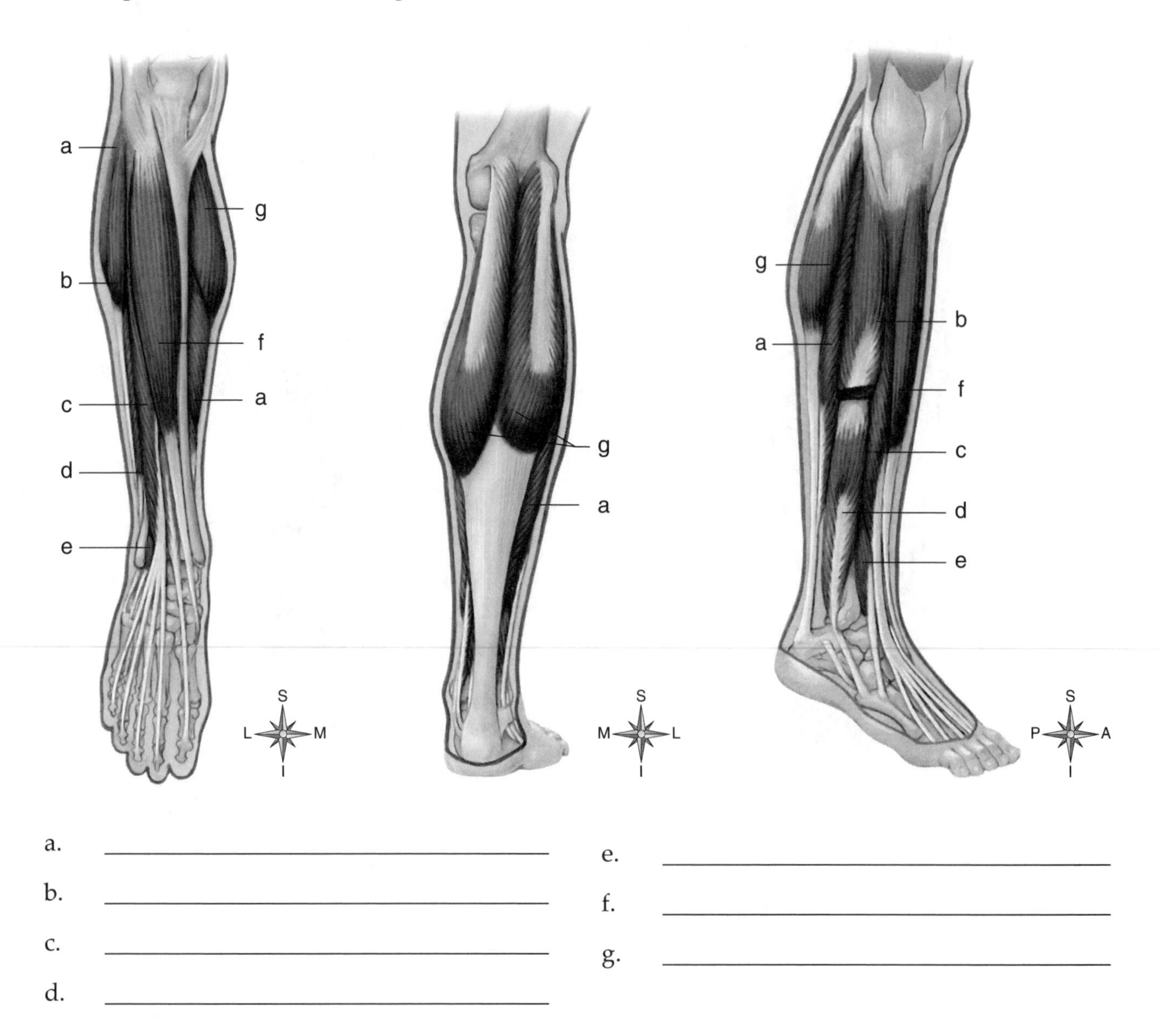

a. ______________________

b. ______________________

c. ______________________

d. ______________________

e. ______________________

f. ______________________

g. ______________________

Short answer

160. Name three muscles that adduct the thigh.

FC 161. Identify the four muscles of the quadriceps femoris group.

FC 162. Identify the three muscles of the hamstrings group.

CHAPTER 11

Cells of the Nervous System

HOW TO APPROACH THE CELLS OF THE NERVOUS SYSTEM

The cells of the nervous system are basically neurons and non-neurons (glial cells). Neurons have only one job: to send, receive, and interpret impulses (INFORMATION). Think of the stereotype of a "nerd"—that is a neuron. Full of knowledge and able to deal with vast amounts of information but completely unable to dress appropriately (insulation), cook, or defend itself.

To transmit and interpret information, you will use ions and move them in and out of the cells (neurons). Once again, a reminder that each chapter builds on the next, so you may need to go back and review chemistry. Focus on what is involved in the conduction of the impulse and how it works. That is the most important component of this chapter and will be helpful to understanding other material in this book and practical applications in life. For example, medications that "sedate" such as sleeping aids, anticonvulsants, and sedatives decrease nerve impulse conduction in the brain by altering influx of ions into the neuron. Active and passive transport (diffusion) are critical in the movement of the ions.

Because neurons ONLY deal with information, they are in need of protection, feeding, insulation, and so on, and that is where glial cells come in. Think of the glial cells as the servants of the neurons. They do everything the neuron cannot.

ORGANIZATION AND CELLS OF THE NERVOUS SYSTEM, NERVE TRACTS

Fill in the blanks with the correct answers.

1. ____________________ are excitable cells that conduct impulses.

2. ____________________ is a sheet of lipids and cholesterol that when wrapped around a nerve fiber is called the ____________________ sheath.

3. The microscopic gaps between Schwann cells are called ____________________.

4. The axon of a neuron is a single fiber that usually extends from a tapered portion of the cell body called the ____________________.

FC 5. A(n) ____________________ is a junction between the synaptic knobs of one neuron and the dendrites of another neuron.

6. Distinct regions of gray matter within the central nervous system (CNS) are called ____________________, but in the peripheral nervous system (PNS) they are called ____________________.

Matching – match each definition to the correct term.

	Term		Definition
7. ______	CNS	a.	"Fight-or-flight"
8. ______	PNS	b.	Voluntary and skeletal muscle
9. ______	Afferent	c.	"Away"
10. ______	Efferent	d.	Brain and spinal cord
11. ______	Somatic nervous system	e.	"Rest-and-repair"
12. ______	Autonomic nervous system	f.	Spinal nerves
FC 13. ______	Sympathetic nervous system	g.	"Toward"
FC 14. ______	Parasympathetic nervous system	h.	Involuntary and visceral

Matching – match each definition to the correct term.

	Term		Definition
15. ______	Astrocytes	a.	Tapered portion of cell body
16. ______	Microglia	b.	Surrounds each axon
17. ______	Ependymal cells	c.	Produces myelin in CNS
18. ______	Oligodendrocytes	d.	Produces cerebrospinal fluid (CSF)
19. ______	Schwann cells	e.	"Protection"
20. ______	Endoneurium	f.	Holds fascicles together
21. ______	Perineurium	g.	Form the blood-brain barrier (BBB)
22. ______	Epineurium	h.	Produces myelin in PNS
23. ______	Axon hillock	i.	Surrounds entire nerve

Multiple choice – select the best answer.

24. Nerve impulses going toward the brain travel along what pathways?
   a. Efferent
   b. Motor
   c. Proximal
   d. Afferent

FC 25. Which of the following types of cells supports and protects neurons?
   a. Glial
   b. Neuronocytes
   c. Glialcytes
   d. Dendrites

26. Information going to the muscles travels down what nervous system?
    a. Autonomic
    b. Central
    c. Somatic
    d. Sympathetic

FC 27. The "fight-or-flight" response is associated with which of the following?
    a. Sympathetic
    b. Parasympathetic
    c. Autonomic
    d. Somatic

28. Which of the following cells "feeds" the neurons by extracting glucose from the blood?
    a. Microglia
    b. Ependymal
    c. Schwann
    d. Astrocyte

29. Which of the following cells forms the BBB?
    a. Astrocytes
    b. Oligodendrocytes
    c. Schwann
    d. Microglia

30. Which of the following cells are only found in the peripheral nervous system?
    a. Ependymal
    b. Schwann
    c. Oligodendrocytes
    d. Microglia

31. These cells resemble epithelial cells and are responsible for producing fluid in the CNS.
    a. Schwann
    b. Ependymal
    c. Microglia
    d. Astrocytes

32. The axon terminates at the:
    a. dendrite.
    b. synaptic knob.
    c. soma.
    d. endoneurium.

FC 33. Which of the following is responsible for transmission of a nerve signal from one neuron to another?
    a. Neurotransmitters
    b. Schwann cells
    c. Synaptic transmitters
    d. Neuroenzymes

34. Bundles of nerve fibers (axons) in the PNS are called:
    a. tracts.
    b. fascicles.
    c. multipolar.
    d. afferent.

35. Which of the following contains large quantities of myelinated fibers?
    a. Gray matter
    b. Nuclei
    c. Ganglia
    d. White matter

STUDY TIP

Think of a nerve cell like your computer. The soma is the tower or hard drive. The keyboard, mouse, and any other sensory devices (taking information toward the computer to be processed) are dendrites which carry information to (afferent) the soma. The information is processed in the computer (soma) and can be sent to another computer via the Internet. The computer connects to other computers by long, insulated wires (axons) carrying information away (efferent) from your computer.

Labeling – label the following diagram.

36. Structure of a typical neuron

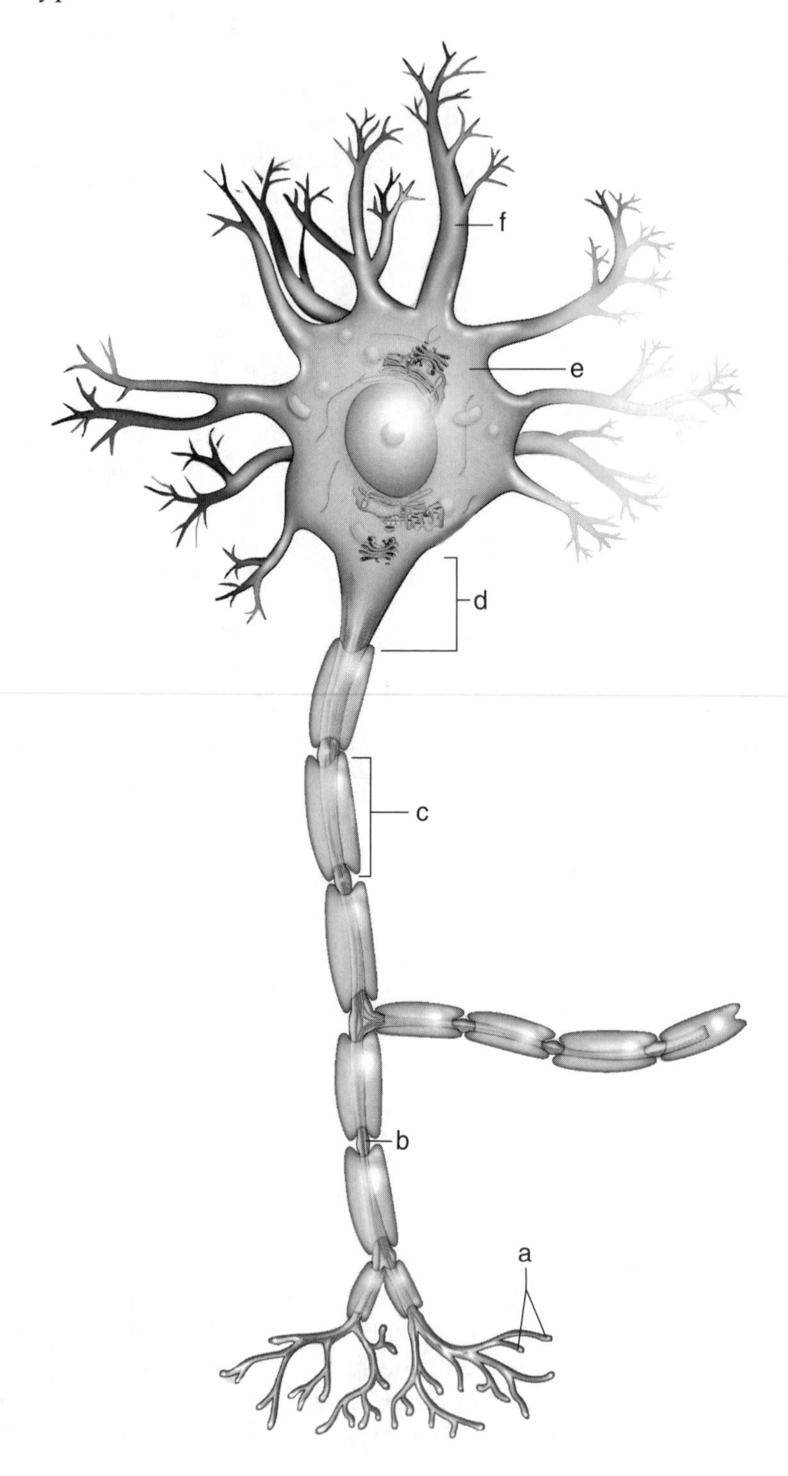

a. ______________________

b. ______________________

c. ______________________

d. ______________________

e. ______________________

f. ______________________

Short answer

FC 37. List the five basic types of glia.

38. List three distinct structural types of neurons.

FC 39. List three types of neurons classified by direction of impulse.

40. Identify the three connective tissue coverings holding together nerve bundles (list them profundus to superficial).

NERVE IMPULSES, ACTION POTENTIAL, AND SYNAPSES

Fill in the blanks with the correct answers.

FC 41. A(n) ____________________ is a wave of electrical energy that travels along the plasma membrane of the nerve.

FC 42. The more ____________________ ions are located outside the neuron and the more ____________________ ions inside the neuron.

43. The ____________________ the diameter of the nerve fiber, the faster it conducts impulses.

FC 44. A(n) ____________________ is the place where signals are transmitted from one neuron to another.

45. Many ____________________ work by inhibiting the opening of sodium channels in the nerve cell.

Matching – match each description, word, or phrase to the correct term.

	Term		Definition
46. ______	Nerve impulse	a.	Potential voltage difference, usually about –70 mV
47. ______	Saltatory conduction	b.	Exceeding this opens sodium channels
48. ______	Resting membrane potential	c.	Wave of electrical energy traveling along a plasma membrane
49. ______	Synapse	d.	Sodium moving in
50. ______	Action potential	e.	"Jumping" from node to node
51. ______	Threshold potential	f.	"Nerve impulse"
FC 52. ______	Repolarization	g.	Synaptic cleft, presynaptic and postsynaptic neuron
FC 53. ______	Depolarization	h.	Positive outside, negative inside
FC 54. ______	Polarized	i.	Potassium moving out

Multiple choice – select the best answer.

55. The normal state that results in a difference in electrical charge across the plasma membranes is called:
   a. cell membrane potential.
   b. impulse.
   c. depolarized.
   d. voltage potential.

FC 56. What is the number-one intracellular ion?
   a. Sodium
   b. Potassium
   c. Calcium
   d. Chloride

57. Which of the following is happening during depolarization?
   a. Sodium is rushing into the cell.
   b. Sodium is rushing out of the cell.
   c. Potassium is rushing into the cell.
   d. Potassium is rushing out of the cell.

FC 58. What is the number-one extracellular ion?
   a. Sodium
   b. Potassium
   c. Calcium
   d. Chloride

59. The membrane potential is measured in:
   a. amps.
   b. megavolts.
   c. microvolts.
   d. millivolts.

60. The movement of the membrane potential toward zero is called:
   a. polarization.
   b. depolarization.
   c. repolarization.
   d. prepolarization.

61. What is it called when the action potential jumps between gaps in the myelin sheath?
   a. Skip conduction
   b. Saltatory conduction
   c. Nodal conduction
   d. Myelination

FC 62. Which of the following is happening during repolarization?
   a. Sodium rushes into the cell.
   b. Calcium rushes out of the cell.
   c. Potassium rushes out of the cell.
   d. Sodium rushes out of the cell.

63. What happens when the threshold potential (about –59 mV) is reached?
   a. All of the sodium channels open.
   b. All of the sodium channels close.
   c. Some of the sodium channels open.
   d. Some of the sodium channels close.

64. What is it called when the membrane will not respond to any other stimulus, no matter how strong?
   a. Repolarization
   b. Saltatory conduction
   c. Synaptic pause
   d. Absolute refractory period

FC 65. The neuron that is releasing the neurotransmitter is called the:
   a. presynaptic neuron.
   b. synaptic neuron.
   c. postsynaptic neuron.
   d. afferent neuron.

66. Which of the following is true regarding electrical synapses?
   a. Require higher concentrations of sodium
   b. Require lower concentrations of neurotransmitters
   c. Do not require neurotransmitters
   d. Have fewer synaptic clefts

67. Which of the following happens when a neurotransmitter binds to a receptor site?
   a. Causes release of more neurotransmitters
   b. Opens sodium channels
   c. Opens potassium channels
   d. Releases acetylcholine

FC 68. Acetylcholine crosses the:
   a. synapse.
   b. synaptic knob.
   c. synaptic cleft.
   d. postsynaptic gap.

Labeling – label the following diagram.

69. Nerve impulse conduction across the synapse

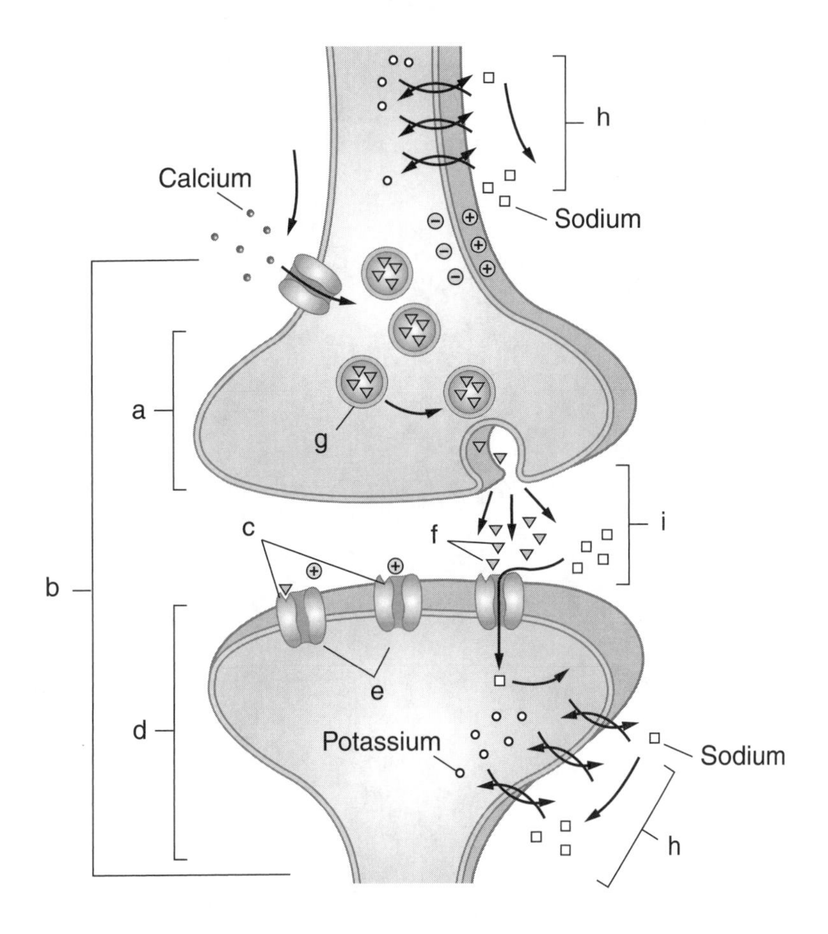

a. ______________________

b. ______________________

c. ______________________

d. ______________________

e. ______________________

f. ______________________

g. ______________________

h. ______________________

i. ______________________

CHAPTER 12

Central Nervous System

HOW TO APPROACH THE CNS

Start this chapter with learning the coverings (meninges) of the brain outside to in or superficial to profundus. They are the same coverings of the spinal cord. Then focus on making sure you understand the difference between gray and white matter and their function. The gray matter is for higher brain functions like thinking, processing, dreaming, learning, and creativity. The white matter is mostly "wiring," and other than carrying signals, has very simple functions such as regulation of such things as vital signs and reflexes. Keep this in mind when you look at pictures of the brain.

With the above foundation, you should spend the majority of the time learning the structures of the central nervous system (CNS) and their function(s). A lot of this chapter is simply memorizing the particular structures and their functions. Flash cards can be very beneficial in this task.

INTRODUCTION, COVERINGS OF THE BRAIN AND SPINAL CORD, CEREBROSPINAL FLUID, AND SPINAL CORD

Fill in the blanks with the correct answers.

1. The ____________________ provides conduction routes to and from the brain.

FC 2. The ____________________ root carries information into the spinal cord and the ____________________ root carries information out of the spinal cord.

3. The spinal cord ends at vertebra ____________________.

4. When the spinal cord ends, it branches out into a "horse tail" of spinal nerve roots called the ____________________.

5. The gray matter of the spinal cord is broken up into gray ____________________ and the white matter is broken up into white ____________________.

FC 6. ____________________ tracts conduct sensory impulses to the brain and ____________________ tracts conduct motor impulses from the brain.

7. The ____________________ and the ____________________ nearly divide the spinal cord into separate, symmetrical halves.

FC *Matching – match each term with the correct description.*

	Term		Definition
8. ______	Meninges	a.	White columns
9. ______	Falx cerebri	b.	Has homeostatic functions similar to blood
10. ______	Tentorium cerebelli	c.	Spinal nerves below L1
11. ______	Cerebrospinal fluid (CSF)	d.	Membranes covering the brain
12. ______	Choroid plexuses	e.	Motor neurons
13. ______	Dorsal root	f.	Produces CSF
14. ______	Ventral root	g.	Partition between the two hemispheres of the brain
15. ______	Spinal nerve	h.	Separates the cerebellum from the cerebrum
16. ______	Cauda equina	i.	Sensory neurons
17. ______	Funiculus	j.	Dorsal and ventral roots together

Multiple choice – select the best answer.

18. Which meningeal layer is in direct contact with the skull?
    a. Osseous mater
    b. Pia mater
    c. Arachnoid mater
    d. Dura mater

19. Which of the following projects into the longitudinal fissure?
    a. Falx cerebelli
    b. Falx cerebri
    c. Tentorium cerebelli
    d. Tentorium cerebri

20. What meningeal layer is responsible for tentorium cerebelli?
    a. Dura mater
    b. Cerebellar mater
    c. Arachnoid mater
    d. Pia mater

21. Epidural space lies between what two structures?
    a. Dura and arachnoid mater
    b. Pia mater and brain
    c. Pia and arachnoid mater
    d. Bone and dura mater

FC 22. CSF circulates in the:
    a. epidural space.
    b. subdural space.
    c. subarachnoid space.
    d. cerebrospinal space.

FC 23. Fluid-filled canals and cavities within the brain are called:
    a. ventricles.
    b. cerebral cavities.
    c. CSF cavities.
    d. dural cavities.

FC 24. A network of capillaries and ependymal cells that produce CSF is called:
    a. cerebrospinal plexuses.
    b. choroid plexuses.
    c. ventricles.
    d. arachnoid granulations.

25. A lumbar puncture or "spinal tap" is performed by taking CSF from where in the spinal cord?
    a. Ventricles
    b. Choroid plexuses
    c. Epidural space
    d. Subarachnoid space

26. The combination of the dorsal and ventral roots creates a mixed nerve called the:
    a. afferent spinal nerve.
    b. spinal nerve.
    c. spinal tract.
    d. cauda equina.

27. Each white column consists of large bundles of nerve fibers divided into smaller bundles called:
    a. spinal tracts.
    b. funiculi.
    c. horns.
    d. spinal nerves.

28. Which ventricle is directly medial and inferior to the lateral ventricle?
    a. Medial ventricle
    b. Fourth ventricle
    c. Third ventricle
    d. Cerebral aqueduct

Labeling – label the following diagrams.

29. Coverings of the spinal cord

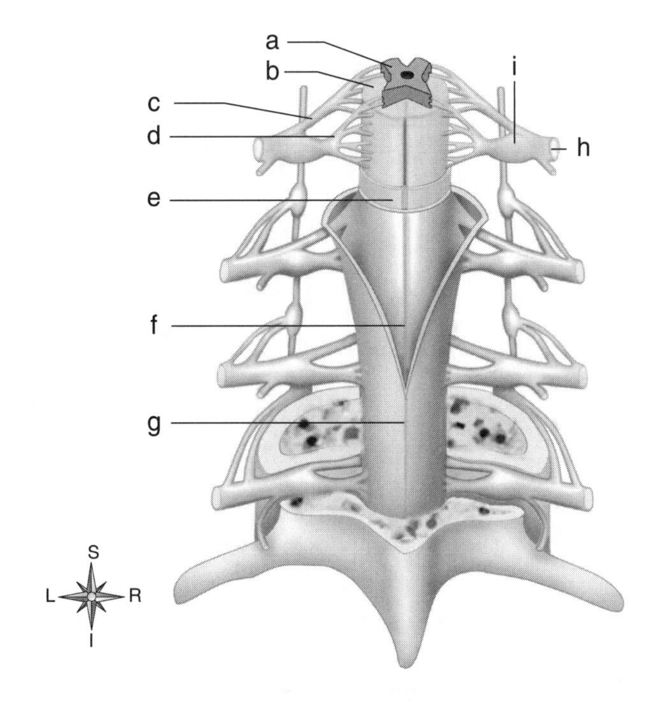

a. ______________________

b. ______________________

c. ______________________

d. ______________________

e. ______________________

f. ______________________

g. ______________________

h. ______________________

i. ______________________

30. Fluid spaces of the brain

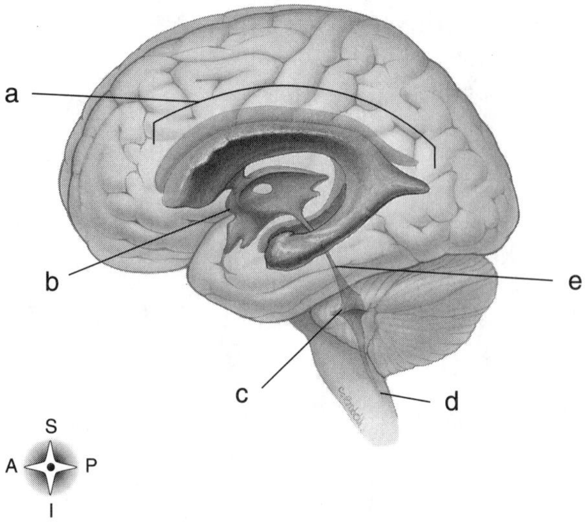

a. ______________________________

b. ______________________________

c. ______________________________

d. ______________________________

e. ______________________________

Short answer

FC 31. List the three distinct meningeal layers superficial to profundus.

32. Identify three important inward extensions of the dura mater.

33. List three important spaces between and around the meninges.

BRAIN AND DIENCEPHALON

Fill in the blanks with the correct answers.

34. Islands of gray matter lying deep inside white matter are called ____________________.

35. Recordings of brain electrical potentials are called ____________________.

36. Without continual stimulation of the cortical neurons (cerebral cortex) by the ____________________, an individual is unconscious and cannot be aroused.

37. The ____________________ is a dumbbell-shaped mass of gray matter made up of several nuclei located in the diencephalon.

38. Two ropelike masses of white matter named ____________________ extend through the midbrain and conduct impulses between the midbrain and the cerebrum.

FC *Matching – match each description with the correct term.*

	Term		Definition
39. ______	Melatonin	a.	Enlargement of upper spinal cord located just above foramen magnum
40. ______	Hypothalamus	b.	White matter of the cerebellum
41. ______	Pons	c.	Mostly a relay station for sensory impulses going to cerebral cortex
42. ______	Infundibulum	d.	Auditory and visual centers
43. ______	Thalamus	e.	Produces melatonin
44. ______	Diencephalon	f.	Hormone that controls "biological clock"
45. ______	Arbor vitae	g.	Controls respiration
46. ______	Corpora quadrigemina	h.	Connects hypothalamus to pituitary gland
47. ______	Medulla oblongata	i.	Regulates hormones, appetite, and temperature
48. ______	Pineal gland	j.	"Between brain" located between cerebrum and midbrain

Multiple choice – select the best answer.

49. Which of the following is NOT part of the brainstem?
    a. Medulla oblongata
    b. Pons
    c. Diencephalon
    d. Midbrain

FC 50. Coordinating skeletal muscle activity is a function of the:
    a. diencephalon.
    b. thalamus.
    c. cerebrum.
    d. cerebellum.

51. There are two bulges of white matter located on the ventral surface of the medulla called:
    a. pyramids.
    b. ventral processes.
    c. anterior processes.
    d. cerebral peduncles.

FC 52. Thermoregulation is controlled where?
    a. Thalamus
    b. Hypothalamus
    c. Infundibulum
    d. Pons

FC 53. Cardiac, respiratory, and vasomotor control centers are located in which of the following?
    a. Pons
    b. Hypothalamus
    c. Nuclei of the medulla
    d. Midbrain

54. Another structure of the midbrain that has auditory and visual reflexes is the:
    a. cerebral peduncles.
    b. corpora quadrigemina.
    c. pineal gland.
    d. pons.

55. The right and left hemispheres of the cerebellum are separated by the:
    a. vermis.
    b. folia.
    c. arbor vitae.
    d. falx cerebri.

56. One of the main functions of the thalamus is:
    a. processing or filtering information going to the cerebrum.
    b. control of vital functions such as heart rate and respiratory rate.
    c. control of body temperature.
    d. production and storage of some hormones.

57. Which of the following is located in the hypothalamus and is responsible for hunger and involved in smell?
    a. Infundibulum
    b. Pineal gland
    c. Corpora quadrigemina
    d. Mamillary bodies

58. Which of the following is a stalk-like attachment leading to the pituitary gland?
    a. Infundibulum
    b. Vermis
    c. Folia
    d. Pituitary chiasma

Labeling – label the following diagram.

59. Divisions of the brain

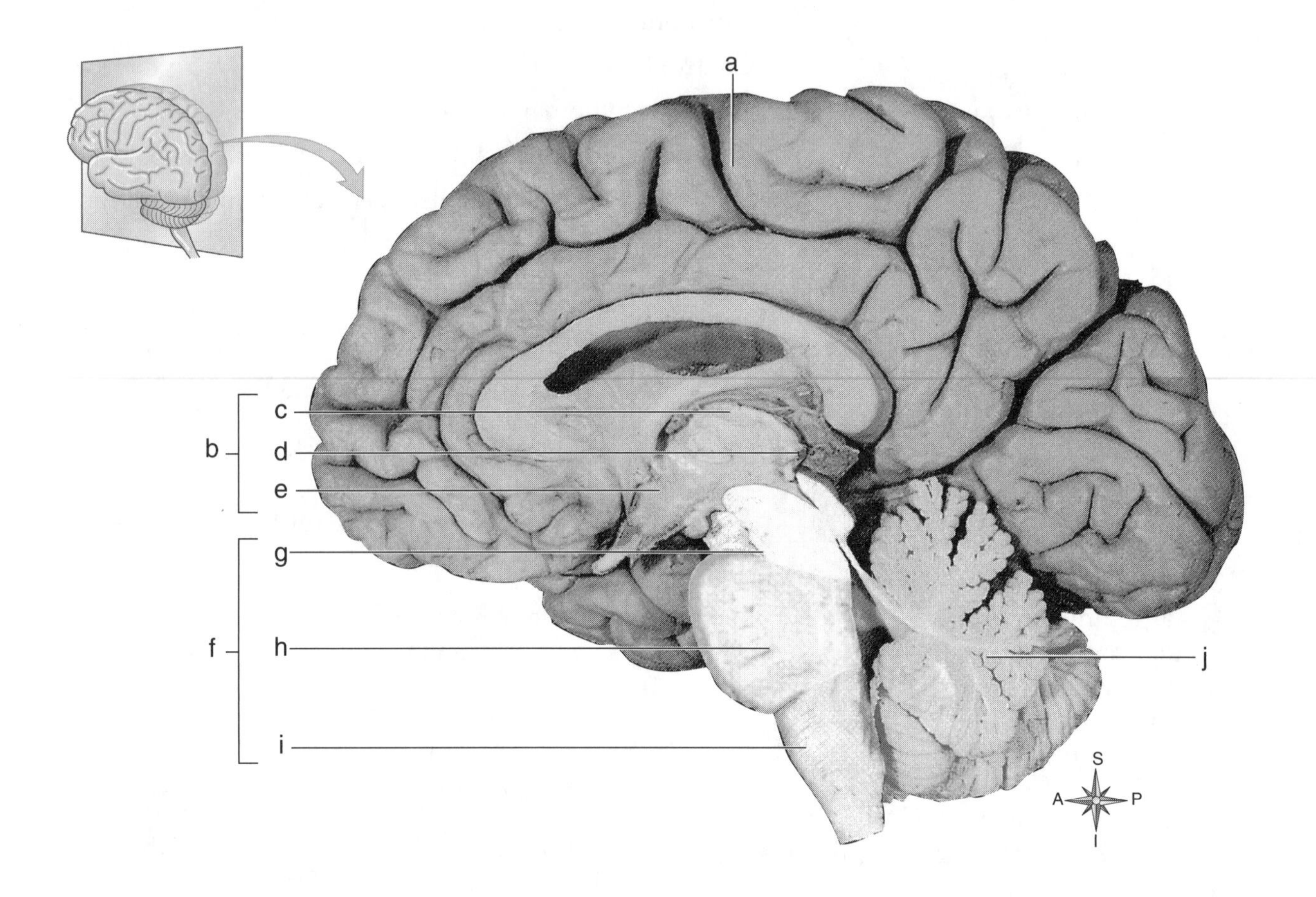

a. ______________________

b. ______________________

c. ______________________

d. ______________________

e. ______________________

f. ______________________

g. ______________________

h. ______________________

i. ______________________

j. ______________________

Short answer

FC 60. Name the five lobes of the brain.

61. Identify the six major divisions of the brain, beginning with the base and working up.

FC 62. Collectively, what three structures make up the brainstem?

STRUCTURES OF THE CEREBRUM AND PATHWAYS OF THE CNS

Fill in the blanks with the correct answers.

63. ____________________ is believed to consist of some kind of structural changes called *engrams* in the cerebral cortex.

FC 64. The surface of the brain is covered with shallow grooves called ____________________ and deeper grooves called ____________________.

65. ____________________ is considered the "speech center" of the brain.

66. ____________________ conduct impulses from the periphery of our bodies to the CNS where the ____________________ consist of motor neurons that conduct impulses from the CNS to the body.

HINT

STUDY TIP

The convolutions of the brain are composed of gyri and sulci. The gyri are like the hills or elevations between the sulci, or valleys or grooves. Just remember "HILLS and VALLEYS."

FC ***Matching – match each description with the correct term.***

	Term		Definition
67. ______	Cerebrum	a.	"White matter"
68. ______	Cerebral cortex	b.	"Rage and fear"
69. ______	Gyrus	c.	Consists of five lobes
70. ______	Sulcus	d.	Regulates voluntary motor functions
71. ______	Corpus callosum	e.	Long-term memory
72. ______	Basal nuclei	f.	"Hills" or elevations
73. ______	Precentral gyrus	g.	"Valleys" or grooves
74. ______	Postcentral gyrus	h.	Primary somatic sensory cortex
75. ______	Limbic system	i.	Outer layer of cerebrum
76. ______	Engrams	j.	Primary somatic motor cortex

Multiple choice – select the best answer.

77. What connects the cerebral hemispheres?
    a. Basal nuclei
    b. Precentral gyrus
    c. Falx cerebri
    d. Corpus callosum

78. What is the thin layer that makes up the surface of the cerebral hemispheres?
    a. Convolutions
    b. White matter
    c. Gray matter
    d. Cerebral cortex

79. Which of the following structures regulates voluntary motor functions such as maintaining posture, walking, and performing other gross or repetitive movements?
    a. Basal nuclei
    b. Corpus callosum
    c. Postcentral gyrus
    d. Occipital lobe

80. The primary somatic motor area is the:
    a. postcentral sulci.
    b. temporal lobe.
    c. occipital lobe.
    d. precentral gyrus.

81. Which of the following increase(s) the surface area of the cerebrum?
    a. Convolutions
    b. Lobes
    c. Fissures
    d. Cerebral cortex

82. Shallow grooves that lie between the convolutions are called:
    a. syrus.
    b. gyri.
    c. sulci.
    d. fissures.

83. Which of the following is considered the primary sensory cortex?
    a. Precentral sulci
    b. Central sulcus
    c. Postcentral gyrus
    d. Cingulated gyrus

84. Which of the following receives impulses from the spinal cord and relays them to the thalamus?
    a. Reticular activating system (RAS)
    b. Broca's area
    c. Corpus callosum
    d. Limbic system

85. What separates frontal from parietal lobes?
    a. Longitudinal fissure
    b. Lateral fissure
    c. Central sulcus
    d. Lateral sulcus

86. Which of the following is NOT a lobe of the cerebrum?
    a. Insula
    b. Limbic
    c. Parietal
    d. Occipital

87. Which of the following travel up the spinal cord, through the brainstem, and terminate in the thalamus?
    a. Motor pathways
    b. Somatic motor pathways
    c. Secondary sensory neurons
    d. Primary sensory neurons

Labeling – label the following diagrams.

88. Left hemisphere of cerebrum, lateral surface

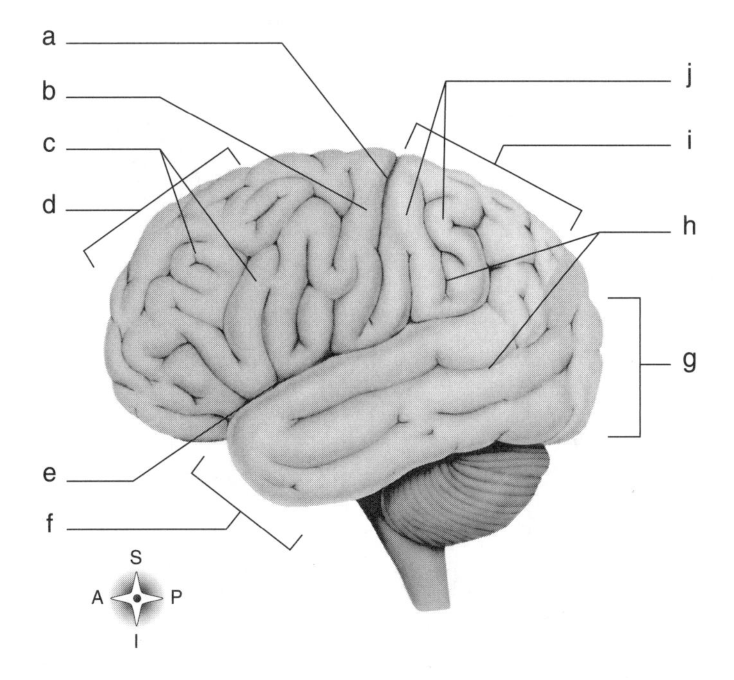

a. ______________________ f. ______________________

b. ______________________ g. ______________________

c. ______________________ h. ______________________

d. ______________________ i. ______________________

e. ______________________ j. ______________________

STUDY TIP

If it is going TO the brain, it is a sensory and/or afferent nerve pathway. If it is coming FROM the brain, it is a motor and/or efferent nerve pathway. Just remember:

SENSORY = TO and MOTOR = FROM

SENSORY = TO = AFFERENT = ASCENDING

MOTOR = FROM = EFFERENT = DESCENDING

CHAPTER 13

Peripheral Nervous System

HOW TO APPROACH THE PNS

The three major topics in this chapter are peripheral nerves/plexuses, cranial nerves, and the autonomic nervous system (ANS). Know what nerves are, how they function, how they form plexuses, and what nerves are involved. Most instructors like to ask a lot of questions about cranial nerves. If you use the study tip below, you should be able to answer almost every question you will encounter. Be sure to know the number, name, function, type (sensory, motor or afferent, efferent or both), possibly how you might test it, and its location.

The ANS is really quite simple. See the study tip below and follow the directions. You will also need to know a few more specifics such as which neurotransmitter is used in both sympathetic and parasympathetic nervous systems and you may need to know some pathways. Focus on what is covered in your class.

SPINAL NERVES

Fill in the blanks with the correct answers.

FC 1. The peripheral nervous system (PNS) is comprised of 31 pairs of ____________________ and 12 pairs of ____________________.

2. There are ____________________ cervical nerve pairs, ____________________ thoracic nerve pairs, ____________________ lumbar nerve pairs, ____________________ sacral nerve pairs, and ____________________ coccygeal pair of spinal nerves.

3. Ventral rami of many spinal nerves subdivide to form braided networks called ____________________.

Matching – match each description with the correct term.

	Term		Definition
4. ______	Spinal nerves	a.	Anterior
5. ______	Ventral root	b.	Soma or gray matter
6. ______	Dorsal root	c.	Motor and sensory
7. ______	Dorsal root ganglion	d.	Ventral rami
8. ______	Dorsal ramus	e.	Formed by dorsal and ventral roots
9. ______	Ventral ramus	f.	Regulates breathing
FC 10. ______	Plexuses	g.	Sciatic nerve
11. ______	Cervical plexus	h.	Posterior
12. ______	Brachial plexus	i.	Autonomic motor fibers
13. ______	Sacral plexus	j.	Innervate lower part of shoulder and arm

Multiple choice – select the best answer.

14. What supplies somatic motor and sensory fibers to several smaller nerves?
    a. Ventral plexus
    b. Dorsal ramus
    c. Dorsal root
    d. Ventral root

FC 15. All thoracic nerves leave the spinal cord through:
    a. vertebral foramina.
    b. spinal foramina.
    c. intervertebral foramina.
    d. transverse foramina.

FC 16. Which of the following contains the cell bodies of the sensory neurons?
    a. Dorsal ramus
    b. Dorsal root ganglion
    c. Ventral root ganglion
    d. Ventral ramus

17. After spinal nerves emerge from the spinal cavity, they form two branches called:
    a. rami.
    b. plexuses.
    c. roots.
    d. ganglia.

18. Which of the following leads to the autonomic motor fibers heading toward the sympathetic chain?
    a. Sympathetic root
    b. Ventral ramus
    c. Dorsal ramus
    d. Motor ramus

19. The phrenic nerve innervates which of the following?
    a. Arm
    b. Arm and shoulder
    c. Diaphragm
    d. Organs

20. The sciatic nerve comes from which plexus?
    a. Lumbar
    b. Brachial
    c. Cervical
    d. Sacral

FC *Short answer*

21. List the four major plexuses.

22. List the spinal nerves that make up the cervical plexus.

23. List the spinal nerves that make up the brachial plexus.

24. List the spinal nerves that make up the lumbar plexus.

25. List the spinal nerves that make up the sacral and coccygeal plexuses.

CRANIAL NERVES

STUDY TIP

FC How to remember the cranial nerves (remember to write this down on exams if possible).

#	Nerve	Mnemonic	Function	Mnemonic
I.	Olfactory	On	Sensory	Some
II.	Optic	Old	Sensory	Suckers
III.	Oculomotor	Olympus'	Motor	May
IV.	Trochlear	Towering	Motor	Marry
V.	Trigeminal	Tops	Both	But
VI.	Abducens	A	Motor	My
VII.	Facial	Fat	Both	Brother
VIII.	Vestibulocochlear	Vicious	Sensory	Says
IX.	Glossopharyngeal	Giant	Both	Be
X.	Vagus	Vaults	Both	Brave
XI.	Accessory	And	Motor	My
XII.	Hypoglossal	Hops	Motor	Man

FC *Matching – match each description with the cranial nerve.*

	Term		Definition
26. ______	Trigeminal	a.	Tongue
27. ______	Trochlear	b.	Trapezius and sternocleidomastoid muscles
28. ______	Abducens	c.	Parasympathetic
29. ______	Vestibulocochlear	d.	Tongue and pharynx
30. ______	Vagus	e.	Hearing and balance
31. ______	Glossopharyngeal	f.	"Smiling"
32. ______	Optic	g.	Lateral rectus muscles
33. ______	Hypoglossal	h.	"Chewing"
34. ______	Oculomotor	i.	Superior oblique muscle
35. ______	Accessory	j.	Intrinsic muscles of the eye
36. ______	Olfactory	k.	"Vision"
37. ______	Facial	l.	"Smell"

STUDY TIP

When memorizing the functions of the cranial nerves (and anything for that matter), keep it as simple as possible. A saying I learned in the military was "keep it simple, stupid." This has served me well. Take the matching sections for example. That is about as simple as it gets. One term and a one- or two-word description. Example: Hypoglossal – tongue. Keep this in mind when making flash cards!

Multiple choice – select the best answer.

38. When you go to the doctor and read the eye chart, which cranial nerve is being tested?
    a. I
    b. II
    c. III
    d. IV

39. What nerve(s) is / are being tested when a light is shined in your eyes to see how your pupils react?
    a. II, III, IV, VI
    b. III
    c. II
    d. II and III

40. If you can look 360 degrees with your eyes, what nerves are being used?
    a. II, III, IV, VI
    b. III and IV
    c. III, IV, and IV
    d. IV and VI

41. Which of the following nerves contains the ophthalmic, maxillary, and mandibular branches?
    a. Vagus
    b. Trigeminal
    c. Abducens
    d. Facial

42. You could test this nerve by having the patient say "light, tight, dynamite."
    a. Glossopharyngeal
    b. Facial
    c. Hypoglossal
    d. Vagus

43. This nerve is responsible for "visceral" activities.
    a. Vagus
    b. Accessory
    c. Trochlear
    d. Hypoglossal

44. Police frequently check this cranial nerve by having the person walk a straight line.
    a. VIII
    b. V
    c. VI
    d. X

45. Which of the following nerves would be tested by having the patient turn her head side-to-side?
    a. III
    b. V
    c. IX
    d. XI

46. Which of the following best describes the optic nerve?
    a. Motor
    b. Mixed
    c. Sensory
    d. Efferent

Labeling – label the following diagram with the appropriate cranial nerve(s).

47. Cranial nerves

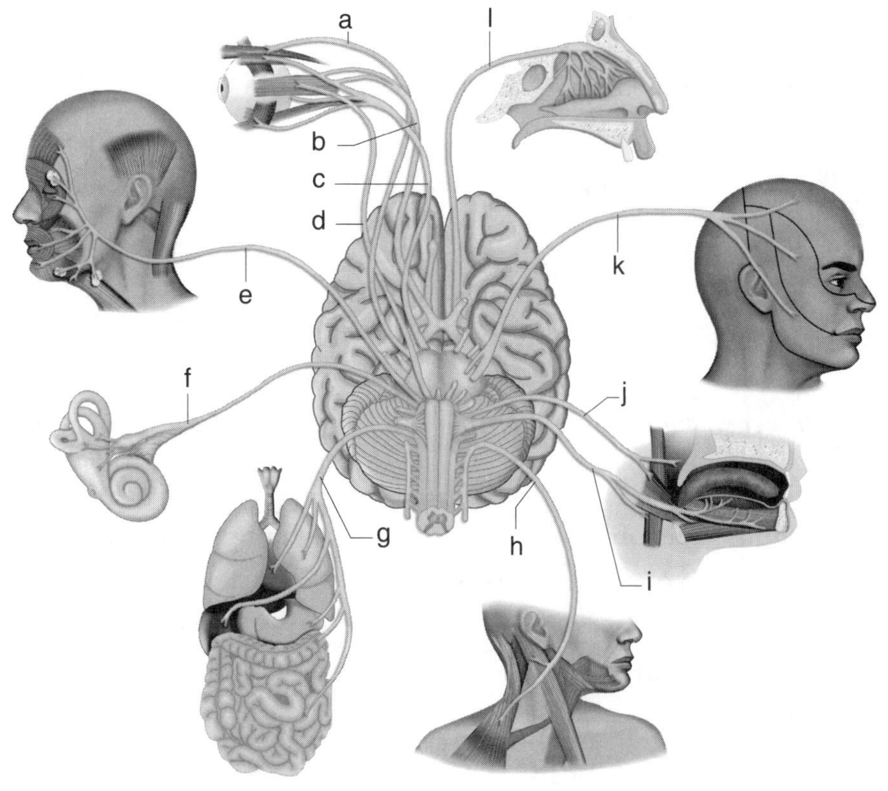

a. ______________________

b. ______________________

c. ______________________

d. ______________________

e. ______________________

f. ______________________

g. ______________________

h. ______________________

i. ______________________

j. ______________________

k. ______________________

l. ______________________

DIVISIONS OF THE PERIPHERAL NERVOUS SYSTEM

STUDY TIP

You can remember the sympathetic and parasympathetic nervous systems and what happens with both just by thinking about the last time you were suddenly really scared. Most people would say they had an "adrenalin dump." Adrenalin is epinephrine and the neurotransmitter that regulates the sympathetic nervous system. How do you feel when you get an adrenalin dump? These are the effects of the sympathetic nervous system: increased heart rate (HR), pallor, increased respiratory rate (RR), increased blood pressure (BP), etc. To remember the parasympathetic nervous system, you simply do the opposite of the systemic (decreased HR, RR, BP, etc.) and remember the neurotransmitter is acetylcholine.

Fill in the blanks with the correct answers.

48. A(n) ____________________ is a predictable response to a stimulus.

49. ____________________ reflexes involve contractions of skeletal muscles.

50. The sympathetic division and parasympathetic division are ____________________ to one another.

51. A(n) ____________________ neuron conducts impulses from the brainstem or spinal cord to an autonomic ganglion.

52. The axon terminals of autonomic neurons release either ____________________ or ____________________.

FC ***Matching – identify each body reaction with an "S" for sympathetic or a "P" for parasympathetic.***

	Body reaction	Autonomic Nervous System
53. ______	Faster contraction of skeletal muscles	Sympathetic
54. ______	Defecation	Parasympathetic
55. ______	Increased production of saliva	
56. ______	Dilated bronchi	
57. ______	Digestion	
58. ______	Stronger heartbeat	
59. ______	Urination	
60. ______	Release of glucose from liver	
61. ______	Increased blood pressure	
62. ______	Decreased respiratory rate	

Multiple choice – select the best answer.

FC 63. Which system is frequently referred to as the "rest and repair" system?
  a. Adrenergic
  b. Anticholinergic
  c. Sympathetic
  d. Parasympathetic

64. What kind of reflex involves contractions of smooth or cardiac muscle?
  a. Somatic
  b. Babinski
  c. Autonomic
  d. Afferent

65. A majority of the parasympathetic preganglionic fibers travel in the __________ before synapsing with postganglionic fibers.
  a. spinal cord
  b. vagus nerve
  c. terminal ganglia
  d. splanchnic nerves

66. Acetylcholine binds with all of the following receptors EXCEPT:
  a. cholinergic.
  b. nicotinic.
  c. muscarinic.
  d. adrenergic.

FC 67. Which of the following is considered the "fight or flight" response?
  a. Parasympathetic
  b. Sympathetic
  c. Cholinergic
  d. Anticholinergic

68. Which of the following is NOT a response of stimulation of the sympathetic nervous system?
  a. Secreting digestive enzymes
  b. Increased blood glucose
  c. Perspiration
  d. Increased blood pressure

69. Alpha and beta receptors are:
  a. cholinergic receptors.
  b. parasympathetic receptors.
  c. adrenergic receptors.
  d. ganglionic receptors.

70. What begins in the gray matter of the spinal cord where their cell bodies and dendrites are located within the thoracic and lumbar segments?
  a. Sympathetic postganglionic neurons
  b. Parasympathetic preganglionic neurons
  c. Sympathetic preganglionic neurons
  d. Parasympathetic postganglionic neurons

71. Which neurons have their cell bodies located in the nuclei of the brainstem?
  a. Sympathetic postganglionic neurons
  b. Parasympathetic preganglionic neurons
  c. Sympathetic preganglionic neurons
  d. Parasympathetic postganglionic neurons

FC 72. Axons that release norepinephrine are called:
  a. cholinergic fibers.
  b. adrenergic fibers.
  c. nicotinic fibers.
  d. muscarinic fibers.

FC 73. Which of the following is the neurotransmitter of the parasympathetic system?
  a. Norepinephrine
  b. Epinephrine
  c. Acetylcholinesterase
  d. Acetylcholine

Short answer

FC 74. What are the two functional divisions of the peripheral nervous system?

75. List four common reflexes often tested by physicians.

FC 76. The ANS is subdivided into what divisions?

FC 77. List five effects of the sympathetic pathway.

FC 78. List five effects of the parasympathetic pathway.

CHAPTER 14

Sense Organs

HOW TO APPROACH THE SENSE ORGANS

Before learning about the special senses, you will need to master the basics. The "basics" are the receptors. Be able to list all the types of receptors and what they sense, where they are, and how they work. See the study tip below for the types of receptors. After you have a good understanding of them, move to the special senses. My experience is that a majority of instructors will focus on sense of sight and hearing, so pay particular attention to these two. One way to test yourself is to follow the path of light or sound and indentify every structure that is passed through or by. If you can do this, you have a good foundation. After naming all the anatomic structures, explain how each works. The process of hearing tends to be more difficult for students to learn, so you may want to spend a little more time on it. Be familiar with taste and smell, but focus on what is covered in class.

RECEPTORS, SENSE OF SMELL AND TASTE

STUDY TIP

How to remember the receptors based on stimuli.

Can	**C**hemoreceptor
The	**T**hermoreceptor
Nurse	**N**ociceptor
Medicate	**M**echanoreceptor
Only	**O**smoreceptor
Patients	**P**hotoreceptor

If you don't like the mnemonic, try making up your own.

Fill in the blanks with the correct answers.

1. Each type of ____________________ responds to a single type of stimulus.

FC 2. When a receptor potential reaches a threshold, it triggers a particular ____________________ in the brain.

3. ____________________ is being aware of a particular sensation.

4. ____________________ proprioceptors are non-adapting and work without movement while ____________________ proprioceptors are only triggered when there is a change in position.

FC ***Matching – match each receptor with the correct description, word, or location (answers may be used more than once).***

		Definition		**Receptor**
5.	_____	Pain	a.	Mechanoreceptor
6.	_____	Hypothalamus	b.	Chemoreceptor
7.	_____	Light	c.	Thermoreceptor
8.	_____	Does not adapt	d.	Nociceptor
9.	_____	Smell and taste	e.	Photoreceptor
10.	_____	Temperature	f.	Osmoreceptor
11.	_____	Eyes		
12.	_____	Detect changes in concentrations of electrolytes		
13.	_____	Pressure		
14.	_____	Muscles and tendons		

Matching – match each description or word with the correct receptor.

		Receptor		**Definition**
15.	_____	Meissner corpuscle	a.	Deep pressure
16.	_____	Ruffini corpuscle	b.	Taste
17.	_____	Pacini corpuscle	c.	Stretch and vibration
18.	_____	Muscle spindle	d.	Stretching muscle
19.	_____	Golgi tendon	e.	Light touch
20.	_____	Gustatory cells	f.	Muscle contraction

Multiple choice – select the best answer.

21. What must reach a threshold in order to produce a sensation?
    a. Receptor sensation
    b. Proprioceptor
    c. Interoceptor
    d. Receptor potential

22. What is it called when the magnitude of the receptor potential decreases over time in response to a continuous stimulus?
    a. Adaption
    b. Decreased sensation potential
    c. Hypostimulation
    d. Sensory atrophy

FC 23. The general sense organs are usually called:
a. visceral senses.
b. somatic senses.
c. tactile senses.
d. exteroceptor senses.

24. These receptors are located on or very near the surface of the body.
a. Proprioceptors
b. Interoceptors
c. Exteroceptors
d. Dermalceptors

25. Gustatory cells are located:
a. in the nose.
b. on the tongue.
c. in the skin.
d. in the muscles.

FC 26. Which of the following receptors allow us to orient our body in space?
a. Nociceptors
b. Osmoreceptors
c. Proprioceptors
d. Exteroceptors

27. What is the most common, simplest, and most widely distributed sensory receptor?
a. Free nerve endings
b. Encapsulated nerve endings
c. Tactile receptors
d. Stretch receptors

28. Often during a heart attack, the pain is felt in the left shoulder. This is called:
a. somatic pain.
b. referred pain.
c. acute somatic pain.
d. mixed pain.

FC 29. Which of the following picks up sensations of light touch?
a. Pacini corpuscles
b. Ruffini corpuscle
c. Meissner corpuscles
d. Dermal corpuscles

30. Ruffini corpuscles are most sensitive to:
a. light touch.
b. deep pressure.
c. vibration.
d. stretch.

31. Which encapsulated nerve ending detects stretching?
a. Golgi tendon
b. Pacini corpuscle
c. Ruffini corpuscle
d. Muscle spindle

32. Vibration and stretch are sensed by:
a. Pacini corpuscles.
b. muscle spindles.
c. Meissner corpuscles.
d. tactile corpuscles.

33. Excessive muscle contraction stimulates which of the following?
a. Muscle spindles
b. Contractile spindles
c. Golgi tendon organs
d. Muscle fibers

34. "Tastebuds" contain clusters of chemoreceptors called:
a. papillae.
b. gustatory cells.
c. papillary chemoreceptors.
d. tastants.

Short answer

FC 35. List six general senses.

FC 36. List the five special senses.

37. What are the three methods of classifying somatic receptors?

38. What are the three classifications of receptors by location?

FC 39. List the six types of receptors based on the stimuli that activate them.

40. What are the two classifications of receptors based on structure?

41. Name three tactile receptors.

42. List the five primary taste sensations.

43. List the four types of tongue papillae.

SENSE OF HEARING AND BALANCE

Fill in the blanks with the correct answers.

44. The ____________________ is the visible part of your ear.

45. The ____________________ consists of hair cells that project into the endolymph and is responsible for hearing.

46. ____________________ surrounds the membranous labyrinth and fills the space between this membranous tunnel and the bony walls that surround it.

47. The roof of the cochlear duct is the ____________________ membrane and the floor is the ____________________ membrane.

48. Vibrations travel through the auditory ossicles to the ____________________, exerting pressure waves into the perilymph of the ____________________. This starts a "ripple" in the perilymph that is transmitted through the ____________________ membrane to the endolymph inside the duct. The transmission proceeds to the ____________________, where we perceive it as sound.

Matching – match each description or word with the correct term.

		Term
FC 49.	______	Oval window
FC 50.	______	Round window
FC 51.	______	Auditory tube
FC 52.	______	Vestibule
FC 53.	______	Semicircular canals
FC 54.	______	Cochlea
FC 55.	______	Macula
56.	______	Otoliths
57.	______	Crista ampullaris
58.	______	Cupula

	Definition
a.	Utricle and saccule
b.	Dynamic equilibrium
c.	Exits cochlea
d.	Sensory epithelium
e.	Stapes
f.	Contains hair cells in semicircular canals
g.	"Ear stones"
h.	Covers and bends hair cells in semicircular canals
i.	Equalizes pressure
j.	Organ of Corti

Multiple choice – select the best answer.

59. The labyrinth refers to the:
    a. middle ear.
    b. vestibule.
    c. cochlea.
    d. inner ear.

FC 60. What separates the external ear from the middle ear?
    a. Round window
    b. Oval window
    c. Tympanic membrane
    d. Auditory membrane

FC 61. Which of the following equalizes the pressure of the middle ear?
    a. Round window
    b. Auditory tube
    c. Oval window
    d. Tympanic membrane

62. Which of the following is located in the middle ear?
    a. Labyrinth
    b. Cochlea
    c. Vestibule
    d. Auditory ossicles

63. The semicircular canals are involved in:
    a. equalizing pressure.
    b. balance during movement.
    c. sense of position relative to gravity.
    d. hearing.

64. The stapes connects to which of the following?
    a. Oval window
    b. Round window
    c. Malleus
    d. Tympanic membrane

65. What connects the middle ear to the nasopharynx?
    a. Tympanic tube
    b. Auditory tube
    c. Round window
    d. Tympanic canal

66. Which auditory ossicle connects to the tympanic membrane?
    a. Incus
    b. Stapes
    c. Malleus
    d. Cochleus

67. Which of the following is NOT part of the bony labyrinth?
    a. Auditory ossicles
    b. Vestibule
    c. Cochlea
    d. Semicircular canals

68. Endolymph in the semicircular canals moves what when you turn your head?
    a. Otoliths
    b. Cupula
    c. Macula
    d. Saccule

FC 69. The cochlea:
    a. is involved with balance.
    b. is located in the middle ear.
    c. contains the spiral organ.
    d. contains the utricle and saccule.

70. Endolymph is located where?
    a. In the membranous labyrinth
    b. In the utricle
    c. In the middle ear
    d. Surrounds the bony labyrinth

FC 71. Which of the following is involved with hearing?
    a. Utricle
    b. Saccule
    c. Cochlea
    d. Semicircular canals

72. The upper section of the cochlear duct is called the:
    a. scala tympani.
    b. vestibular tympani.
    c. scala vestibuli.
    d. vestibular membrane.

73. What is located in the utricle and is responsible for static equilibrium?
    a. Crista ampullaris
    b. Cupula
    c. Otoliths
    d. Endolymph

74. Cristae ampullaris are located in the:
    a. utricle.
    b. semicircular canals.
    c. cochlea.
    d. saccule.

75. Many people insert cotton swabs into which part of the ear despite the warning on the box?
    a. External acoustic meatus
    b. Auditory tube
    c. Internal auditory canal
    d. Eustachian canal

Labeling – label the following diagrams.

76. The ear

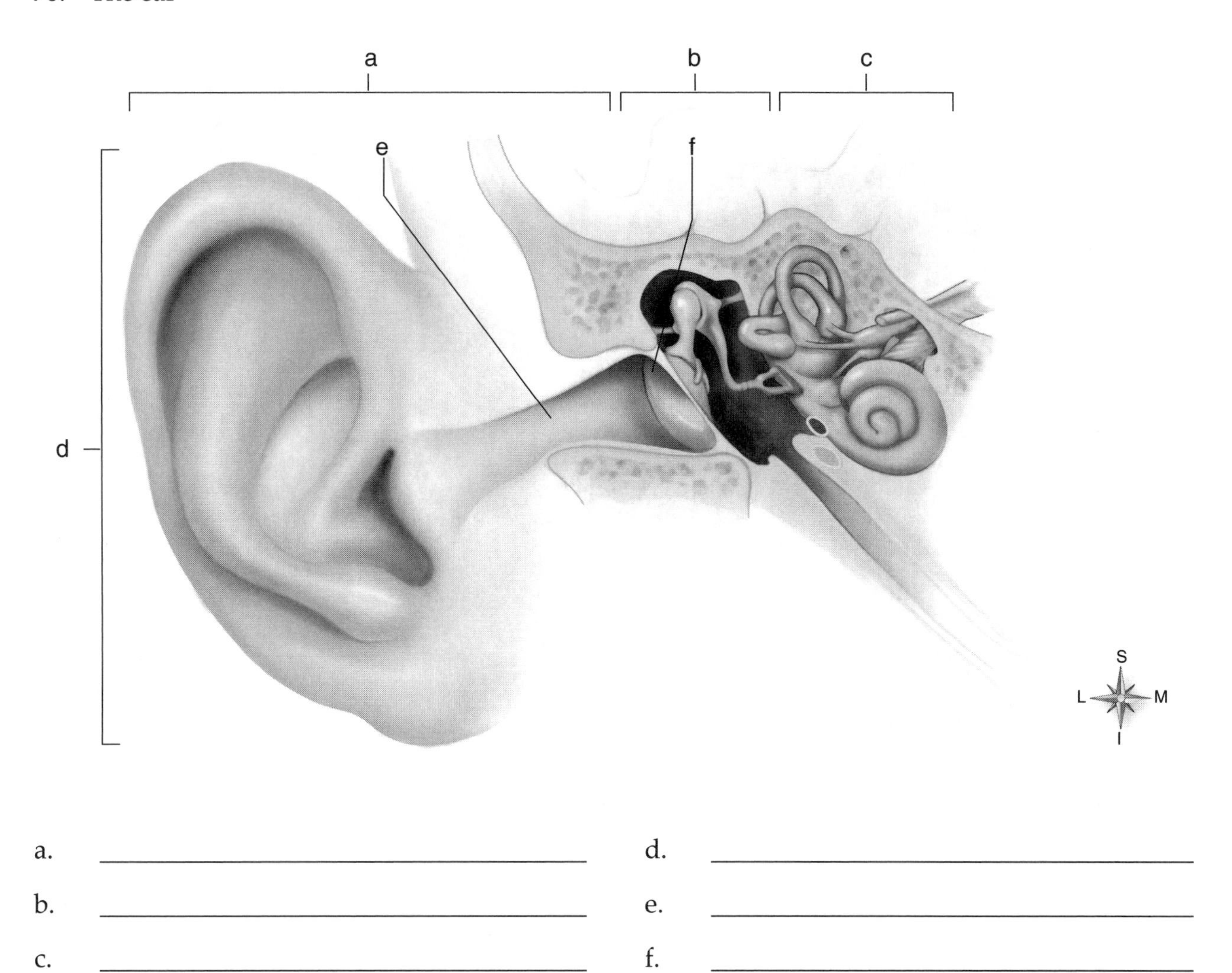

a. ______________________

b. ______________________

c. ______________________

d. ______________________

e. ______________________

f. ______________________

77. Effect of sound waves on cochlear structures

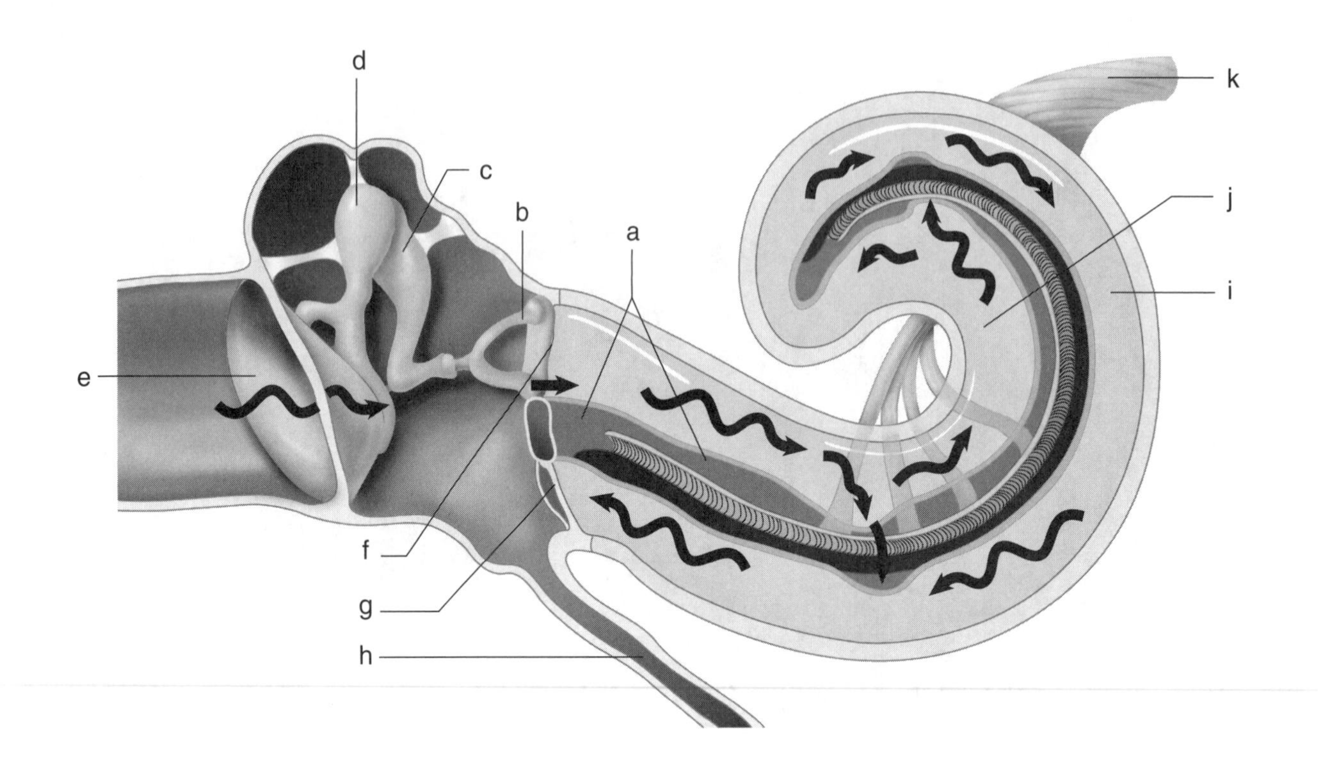

a. ______________________

b. ______________________

c. ______________________

d. ______________________

e. ______________________

f. ______________________

g. ______________________

h. ______________________

i. ______________________

j. ______________________

k. ______________________

FC *Short answer*

78. List the three major anatomical parts or sections of the ear.

79. List the auditory ossicles lateral to medial.

80. The bony labyrinth consists of what three parts?

VISION

Fill in the blanks with the correct answers.

81. A lateral and medial angle or ____________________ forms where the eyelids meet.

FC 82. The ____________________ functions like the diaphragm of a camera, controlling the amount of light entering.

FC 83. The ____________________ is the innermost lining of the eyeball and has no anterior portion.

FC 84. ____________________ detect black and white, while ____________________ detect color.

85. For near vision, the ciliary muscle is ____________________ and the lens is bulging. For far vision, the ciliary muscle is ____________________ and the lens is comparatively flat.

86. All photopigments can be reduced to a glycoprotein called ____________________ and a vitamin A derivative called ____________________.

FC *Matching – match each description or word with the correct term.*

		Term		Definition
87.	______	Convergence	a.	Opening
88.	______	Accommodation	b.	"Whites of the eyes"
89.	______	Refraction	c.	Both eyes come together
90.	______	Blind spot	d.	Convergence
91.	______	Posterior cavity	e.	Bending light
92.	______	Rods	f.	Where a contact lens sits on the eye
93.	______	Pupil	g.	Low light
94.	______	Fovea centralis	h.	Vitreous humor
95.	______	Cornea	i.	Optic disc
96.	______	Sclera	j.	Cones

Multiple choice – select the best answer.

FC 97. What is the correct order of the flow of tears?
   a. Lacrimal gland, lacrimal ducts, lacrimal canals, lacrimal sac, nasolacrimal duct
   b. Lacrimal gland, lacrimal canals, lacrimal ducts, lacrimal sac, nasolacrimal duct
   c. Lacrimal sac, lacrimal gland, lacrimal ducts, lacrimal canal, nasolacrimal canal
   d. Nasolacrimal gland, lacrimal canals, lacrimal sac, lacrimal duct, nasolacrimal canal

FC 98. What is the transparent mucous membrane covering the sclera?
   a. Choroid coat
   b. Punctum
   c. Conjunctiva
   d. Canthus

99. Which of the following is comprised of structures that secrete and drain tears?
   a. Lacrimal ducts
   b. Lacrimal sacs
   c. Lacrimal apparatus
   d. Lacrimal glands

FC 100. Fovea centralis is located in the center of the:
a. optic disc.
b. macula lutea.
c. ciliary body.
d. macular disc.

101. The openings of the lacrimal canals are called:
a. puncta.
b. lacrimal ducts.
c. lacrimal meatuses.
d. lacrimal foramens.

FC 102. The iris regulates the:
a. shape of the lens.
b. size of the pupil.
c. shape of the cornea.
d. ciliary muscle.

FC 103. The suspensory ligaments:
a. hold the eye in place in the orbit.
b. control blinking.
c. connect the ciliary muscle to the lens.
d. pull on the iris, opening or closing the pupil.

FC 104. Which of the following is a transparent portion of the fibrous layer of the eye?
a. Sclera
b. Conjunctiva
c. Choroid
d. Cornea

FC 105. If a person has "blue eyes," what part is colored blue?
a. Iris
b. Cornea
c. Sclera
d. Pupil

FC 106. Rods and cones are located in the:
a. fovea centralis.
b. macula lutea.
c. retina.
d. choroid.

FC 107. The highest concentration of cones is in the:
a. optic disc.
b. macula lutea.
c. periphery of the retina.
d. fovea centralis.

108. Which of the following connects directly to rods and cones?
a. Monopolar neurons
b. Bipolar neurons
c. Ganglion neurons
d. Lateral connecting cells

FC 109. The "blind spot" is the:
a. optic nerve.
b. optic disc.
c. fovea.
d. macula.

110. Aqueous humor comes mostly from capillaries located where?
a. Ciliary body
b. Retina
c. Iris
d. Choroid

111. Which of the following is the correct path of light entering the eye?
a. Cornea, anterior chamber, pupil, posterior chamber, lens, posterior cavity, retina
b. Conjunctiva, cornea, anterior cavity, pupil, lens, posterior cavity, retina
c. Conjunctiva, cornea, posterior cavity, iris, lens, anterior cavity, retina
d. Cornea, anterior cavity, pupil, posterior cavity, lens, posterior chamber, retina

112. In regards to refraction, the more convex the light-bending surface, the:
a. brighter the image will be.
b. greater the refractive power.
c. lesser the refractive power.
d. darker the image will be.

113. Which of the following is NOT required for accommodation for near vision?
    a. Increase in curvature of the lens
    b. Constriction of the pupil
    c. Convergence of the two eyes
    d. Increase in curvature of the cornea

114. What is the only photopigment found in rods?
    a. Rhodopsin
    b. Opsin
    c. Retinal
    d. Rodinal

115. What is the light-absorbing portion of all photopigments?
    a. Vitamin A
    b. Retinal
    c. Opsin
    d. Glycoproteins

116. Bright light is necessary to stimulate which of the following?
    a. Rods
    b. Rhodopsin
    c. Photopigments
    d. Cones

Labeling – label the following diagrams.

117. Lacrimal apparatus

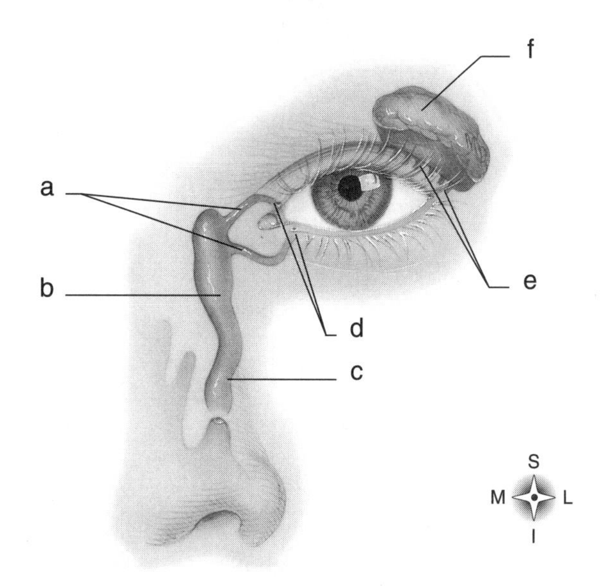

a. ______________________

b. ______________________

c. ______________________

d. ______________________

e. ______________________

f. ______________________

118. Horizontal section through the eyeball

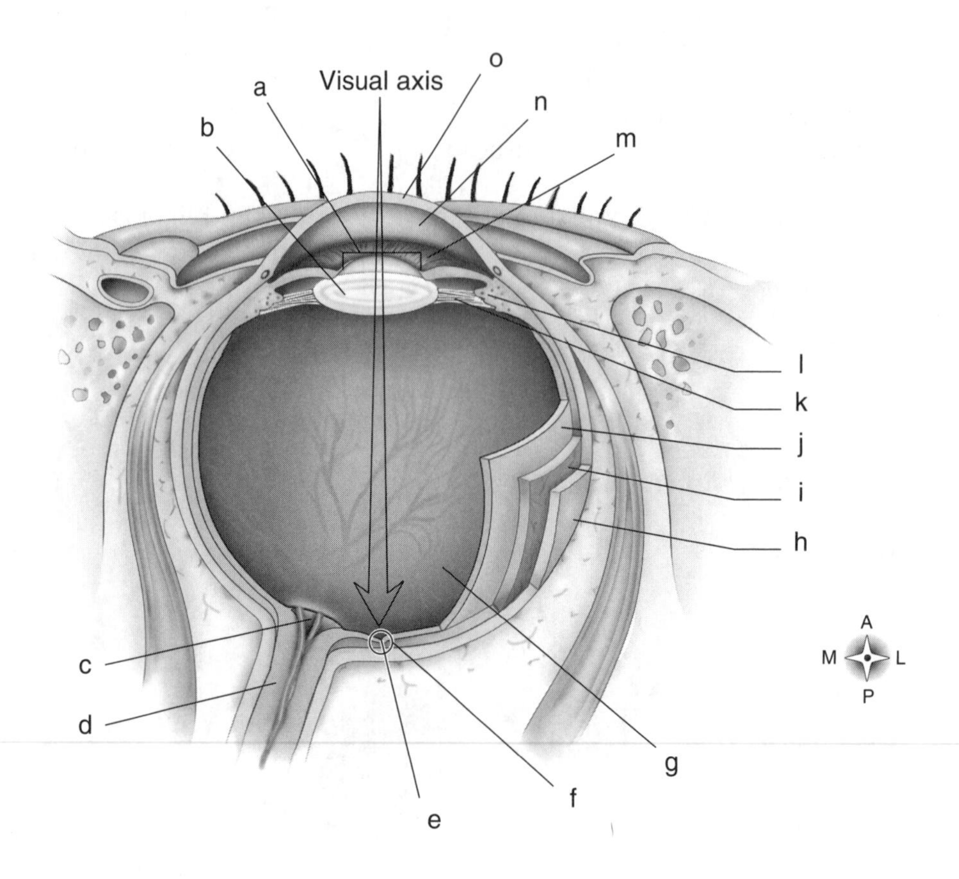

a. ______________________ i. ______________________

b. ______________________ j. ______________________

c. ______________________ k. ______________________

d. ______________________ l. ______________________

e. ______________________ m. ______________________

f. ______________________ n. ______________________

g. ______________________ o. ______________________

h. ______________________

119. Extrinsic muscles of the right eye, lateral view

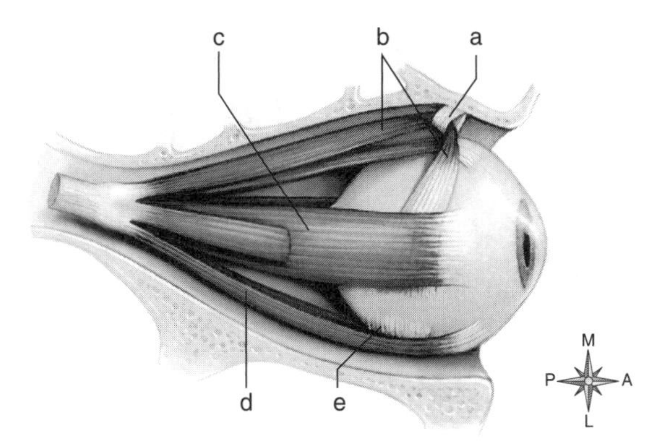

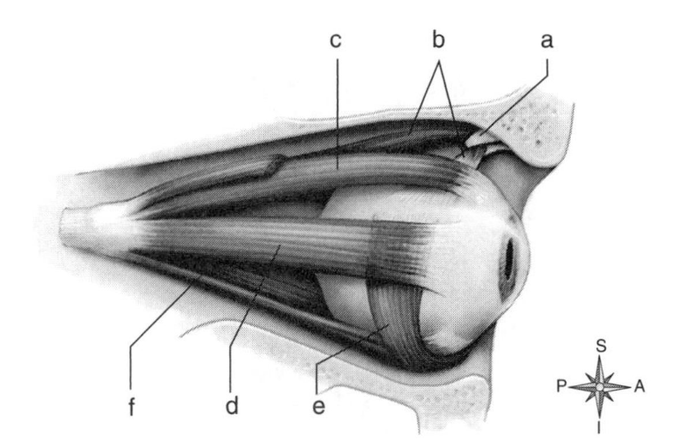

a. ______________________

b. ______________________

c. ______________________

d. ______________________

e. ______________________

f. ______________________

Short answer

120. List four structures of the lacrimal apparatus.

FC 121. Name the six extrinsic muscles of the eye.

FC 122. What are the three layers of the eye?

123. List three components of the vascular layer.

FC 124. Name the two cavities of the eye and the fluid that fills them.

125. Name four refracting media of the eye (what will bend light).

CHAPTER 15

Endocrine System

HOW TO APPROACH THE ENDOCRINE SYSTEM

The focus of this chapter is the glands and the hormones they produce. The key to learning the hormones is first to understand what hormones are and what they do. Next you MUST memorize all the glands, their location, what hormone they produce, the target of the hormone, and the effects of that hormone. Just remember: Hormone, Target, and Action. The tables provided are very useful in this chapter because they have all that information. Make a flash card for each gland and ask the above information with the answers on the back. Remember to keep it simple!

HORMONES AND PROSTAGLANDINS AND RELATED COMPOUNDS

Fill in the blanks with the correct answers.

1. ____________________ signal specific target cells that may exist in a single organ or be distributed throughout the body.

2. We can classify hormones according to their chemical structure as ____________________ or ____________________.

3. Hormones will bind only to ____________________ that "fit" them exactly in the "lock-and-key" mechanism.

4. Any cell with one or more receptors for a particular hormone is a(n) ____________________ of that hormone.

5. A nonsteroid hormone acts as a first messenger, delivering its chemical to the ____________________, where it triggers a series of chemical reactions inside the cell. The message is then passed into the cell where a(n) ____________________ triggers the appropriate cellular changes.

6. The control of hormonal secretion is usually part of a(n) ____________________ loop.

Multiple choice – select the best answer.

7. All of the following are DIFFERENCES between the nervous system and endocrine system EXCEPT:
   a. diffuse into blood.
   b. influence cellular activities.
   c. regulate most of the cells of the body.
   d. have longer effects on the body.

8. What type of hormones target other endocrine glands and stimulate their growth and secretion?
   a. Tropic
   b. Releasing
   c. Activating
   d. Protein

9. Which of the following hormones are manufactured from cholesterol?
   a. Protein hormones
   b. Glycoproteins
   c. Prostaglandins
   d. Steroids

10. What is the complex process where each hormone-receptor interaction produces different regulatory changes within the target cell?
    a. Signal transduction
    b. Synergism
    c. Hormone-receptor activation
    d. Receptor activation

11. The model of signal transduction that steroids produce is called:
    a. first-messenger model.
    b. second-messenger model.
    c. direct-messenger model.
    d. mobile-receptor model.

12. Which types of hormone are not very soluble in the blood and thus must be taken to their target cells by carrier molecules?
    a. Nonsteroidal
    b. Protein
    c. Glycoprotein
    d. Steroids

13. What plays important roles in regulating blood pressure and metabolism and are involved in inflammation and fever?
    a. Tropic hormones
    b. Anabolic hormones
    c. Prostaglandins
    d. Nonsteroid hormones

14. In what model of signal transduction does the hormone easily pass through the plasma membrane?
    a. Mobile-receptor model
    b. Second-messenger model
    c. First-messenger model
    d. Lipid-messenger model

15. What are a unique group of lipids that act as "local hormones"?
    a. Steroids
    b. Prostaglandins
    c. Neuroendocrine hormones
    d. Proximal hormones

Short answer

16. List the four types of nonsteroid hormones.

17. List four things that prostaglandins do.

18. List the two models of signal transduction.

THE GLANDS: PITUITARY, PINEAL, THYROID, PARATHYROID, ADRENAL, PANCREAS, GONADS, PLACENTA, THYMUS, AND OTHER ORGANS AND TISSUES THAT PRODUCE HORMONES

STUDY TIP

You can remember the hormones of the anterior pituitary (or adenohypophysis) by:

Professional	**Pr**olactin
Athletes	**A**CTH
Get	**G**rowth hormone
Fabulous	**F**SH
Lengthy	**L**H
Training	**T**SH

The posterior pituitary (or neurohypophysis) secretes only ADH and oxytocin.

Fill in the blanks with the correct answers.

19. The ____________________ connects the hypothalamus to the pituitary gland.

20. ____________________ cells in the testes produce testosterone.

21. Estrogen is produced by cells of the ovarian ____________________.

22. ____________________ is secreted by the corpus luteum.

STUDY TIP

You may have noticed that all the matching exercises have been flagged for flash cards. I recommend making a flash card for each hormone. Try making the flash cards like the following.

Insulin ***(side A)***
1. Gland –
2. Location –
3. Target –
4. Action –

(side B)
1. Pancreas – alpha cells
2. Peritoneal cavity below and behind stomach (retroperitoneal)
3. Body cells, mostly muscle, liver, and adipose
4. Movement of glucose, amino acids, and fatty acids out of the blood and into cells

FC ***Matching – match the hormone(s) with the gland that produces it (them).***

	Gland		Hormone
23. ______	Anterior pituitary gland	a.	Estrogens and progesterone
24. ______	Posterior pituitary gland	b.	Ghrelin
25. ______	Pineal gland	c.	Calcitonin, thyroxine, and triiodothyronine
26. ______	Thyroid	d.	Melatonin
27. ______	Parathyroid	e.	Atrial natriuretic hormone (ANH)
28. ______	Adrenal	f.	Oxytocin and antidiuretic hormone (ADH)
29. ______	Pancreas	g.	Corticosteroids and catecholamines
30. ______	Testes	h.	Growth hormone (GH), prolactin, thyroid-stimulating hormone (TSH), adrenocorticotropic hormone (ACTH), follicle-stimulating hormone (FSH), and luteinizing hormone (LH)
31. ______	Ovaries	i.	Androgens
32. ______	Placenta	j.	Parathyroid hormone (PTH)
33. ______	Heart	k.	Somatostatin, glucagon, pancreatic polypeptide, and insulin
34. ______	Gastric and intestinal mucosa	l.	Human chorionic gonadotropin (hCG)

FC ***Matching – match the target with the correct hormone (target may be used more than once).***

	Hormone		Target
35. ______	GH	a.	Bone and muscle
36. ______	Prolactin	b.	Hypothalamus
37. ______	TSH	c.	Ovaries
38. ______	ACTH	d.	Breasts
39. ______	FSH	e.	Bone and kidney
40. ______	LH	f.	Bone
41. ______	ADH	g.	Breasts and uterus
42. ______	Oxytocin	h.	Ovary and testes
43. ______	Calcitonin	i.	Cortex of adrenal gland
44. ______	PTH	j.	Thyroid
45. ______	Aldosterone	k.	Uterus
46. ______	Progesterone	l.	Kidney
47. ______	hCG		
48. ______	Ghrelin		
49. ______	ANH		

FC *Matching – match the action with the correct hormone.*

	Hormone		Action
50. ______	ANH	a.	Influences digestion
51. ______	Ghrelin	b.	Stimulates ovaries to secrete progesterone
52. ______	Thymosin	c.	Regulates endocrine cells of the pancreas
53. ______	hCG	d.	Stimulates hypothalamus to increase appetite
54. ______	Progesterone	e.	Growth of muscle and bone
55. ______	Estrogen	f.	Development of immune system
56. ______	Testosterone	g.	Promotes loss of sodium in urine
57. ______	Somatostatin	h.	Maintains the lining of the uterus
58. ______	Pancreatic polypeptide	i.	Stimulates conversion of glycogen to glucose in liver
59. ______	Insulin	j.	Prepares the lining of the uterus
60. ______	Glucagon	k.	Movement of glucose, amino acids, and fatty acids out of the blood and into cells

FC *Matching – match the action with the correct hormone(s).*

	Hormone		Action
61. ______	Glucocorticoids	a.	Contraction of uterus and ejection of milk
62. ______	Aldosterone	b.	Reabsorb calcium from kidney and increase bone breakdown
63. ______	PTH	c.	Breast development and milk production
64. ______	Calcitonin	d.	Gluconeogenesis
65. ______	T_3 and T_4	e.	Increasing sodium reabsorption in kidney
66. ______	Oxytocin	f.	Causes secretion of T_3 and T_4
67. ______	ADH	g.	Increases bone formation and inhibits bone breakdown
68. ______	LH	h.	Regulates metabolism
69. ______	FSH	i.	Secretion of hormones of adrenal cortex
70. ______	ACTH	j.	Promote body growth
71. ______	TSH	k.	Stimulates corpus luteum (ovary) to release progesterone and testes to secrete testosterone
72. ______	GH	l.	Reabsorption of water from kidney
73. ______	Prolactin	m.	Release of egg and development of sperm

STUDY TIP

Remember that estrogens get women ready to get pregnant and progesterone keeps them pregnant!

Multiple choice – select the best answer.

74. Which of the following hormones is released by the anterior pituitary?
    a. Prolactin
    b. TSH
    c. ADH
    d. FSH

75. What carries blood from the hypothalamus directly to the anterior pituitary?
    a. Hypophyseal portal system
    b. Infundibulum
    c. Neurohypophysis
    d. Anterior portal arteries

76. LH and FSH are called:
    a. androgens.
    b. releasing hormones.
    c. gonadotropins.
    d. neurosecretory hormones.

77. What directly regulates the secretions of the anterior pituitary?
    a. Hypothalamus
    b. Neurohypophysis
    c. Pituitary
    d. Infundibulum

78. ADH is also called:
    a. tropic hormone.
    b. vasopressin.
    c. releasing hormone.
    d. somatotropin.

79. Which of the following are nonsteroid hormones called *catecholamines*?
    a. Epinephrine and norepinephrine
    b. ADH and aldosterone
    c. FSH and LH
    d. T_3 and T_4

80. What hormone increases when sunlight is absent?
    a. ADH
    b. T_4
    c. Aldosterone
    d. Melatonin

81. Which of the following is necessary for producing T_3 and T_4?
    a. Calcium
    b. Iodine
    c. Phosphorus
    d. Sodium

82. Which of the following permits calcium to be absorbed through the intestinal tract?
    a. Vitamin D
    b. Vitamin E
    c. Vitamin A
    d. Vitamin C

83. The adrenal cortex produces what hormones?
    a. Mineralocorticoids and glucocorticoids
    b. ADH and aldosterone
    c. Epinephrine and norepinephrine
    d. Catecholamines

84. Aldosterone is a:
    a. mineralocorticoid.
    b. glucocorticoid.
    c. gonadocorticoid.
    d. catecholamine.

85. Which of the following is NOT an action of glucocorticoids?
    a. Accelerate the breakdown of proteins
    b. Essential for maintaining normal blood pressure
    c. Decrease the number of white blood cells
    d. Decrease blood glucose levels

86. The adrenal medulla produces:
    a. catecholamines.
    b. cortisol.
    c. ADH.
    d. corticosteroids.

87. Which of the following hormones is used to test for pregnancy?
    a. hCG
    b. FSH
    c. LH
    d. Progesterone

88. Which gland is in the mediastinum?
    a. Thyroid
    b. Parathyroid
    c. Thymus
    d. Adrenal

STUDY TIP

Mineralocorticoids affect MINERALS, glucocorticoids affect GLUCOSE.

Short answer

89. List the four types of endocrine cells of the pancreatic islets and what they produce.

90. List the six hormones produced by the posterior pituitary gland.

91. List the two hormones produced by the anterior pituitary gland.

92. List the three types of hormones produced by the adrenal glands.

CHAPTER 16

Blood

HOW TO APPROACH THE BLOOD

The key to understanding this chapter is to focus on the formed elements of blood (erythrocytes, thrombocytes, and leukocytes), blood clotting, and dissolving blood clots. You need to be able to recognize all abbreviations such as RBC, WBC, etc. You can quickly test yourself by taking a scratch piece of paper and listing all the formed elements listed above and write down a BRIEF description of function, where they are formed, and what regulates them. You also need to remember what each of the leukocytes responds to. Example: neutrophils – bacteria; eosinophils – parasites, etc. Flash cards are particularly helpful with this. Put the formed element on one side with questions on what you need to remember and the answers on the other side.

The ABO system is simple yet tends to give students difficulty, so make sure you are comfortable with the questions regarding blood transfusions. If you can answer these questions, you are doing well and understand the major concepts. See the study tip in this section. Be sure to use a scratch piece of paper when answering these questions.

For coagulation and fibrinolysis, you should be able to quickly write out on paper the sequence of events for each. Example: fibrinolysis = plasminogen – tPA – plasmin – clot breakdown.

BLOOD COMPOSITION AND RED BLOOD CELLS

Fill in the blanks with the correct answers.

FC 1. ______________________ accounts for 55% of blood, while ______________________ account for 45%.

FC 2. The packed cell volume at the bottom of the test tube is called the ______________________.

3. ______________________ extrudes its nucleus and also loses its ribosomes, mitochondria, and other organelles before the cell reaches maturity in the bone marrow.

4. When four oxygen molecules chemically bond to hemoglobin it is called ____________________ and when carbon dioxide bonds to the hemoglobin it is called ____________________.

FC 5. If oxygen levels decrease, the ____________________ release erythropoietin, which in turn stimulates the ____________________ to increase production of red blood cells (RBCs).

Matching – match each description or word with the correct term.

	Term		Definition
6. ______	Hematopoietic stem cells	a.	Measure of cellular portion of blood
FC 7. ______	Erythropoiesis	b.	Hormone
FC 8. ______	Erythropoietin	c.	Nucleated cells that form erythrocytes
9. ______	Globin	d.	Protein chain
10. ______	Hemoglobin	e.	Liquid portion of blood
11. ______	Erythrocytes	f.	RBCs
12. ______	Hematocrit	g.	Making RBCs
13. ______	Formed elements	h.	Protein
14. ______	Plasma	i.	Cellular portion of blood

Multiple choice – select the best answer.

15. Conditions that result in decreased RBC numbers are called:
    a. hematocrit.
    b. polycythemia.
    c. anemia.
    d. leukopenia.

FC 16. Mature red blood cells are called:
    a. thrombocytes.
    b. erythrocytes.
    c. leukocytes.
    d. hematopoietic cells.

FC 17. Low oxygen levels in the blood increase secretion of what glycoprotein hormone?
    a. Diapedesis
    b. Heparin
    c. Thrombopoietin
    d. Erythropoietin

18. What replaces the nucleus and organelles in mature erythrocytes?
    a. Hemoglobin
    b. Mitochondria
    c. Granules
    d. Histamine

19. Each hemoglobin molecule is composed of _____ protein chains with _____ iron atoms.
    a. 2, 2
    b. 4, 2
    c. 4, 4
    d. 2, 4

20. Which of the following describes erythropoiesis?
    a. Destruction of RBCs
    b. Creation of hematopoietic stem cells
    c. Production of erythropoietin
    d. Process of RBC formation

21. Maturing erythrocytes that have lost their nuclei are called:
    a. EPO.
    b. reticulocytes.
    c. myeloid cells.
    d. megakaryoblasts.

FC 22. Which of the following is the average life span of an RBC?
    a. 30 days
    b. 60 days
    c. 120 days
    d. 160 days

23. Macrophages in what organ(s) primarily recycle RBCs?
    a. Liver
    b. Spleen
    c. Bone marrow, liver, and spleen
    d. Liver and spleen

FC 24. What results from the breakdown of hemoglobin and is/are excreted as part of bile?
    a. Bilirubin
    b. Cholesterol
    c. Iron
    d. Iron and globin

WHITE BLOOD CELLS

STUDY TIP

How to remember the leukocytes and their abundance in the blood:

Nurse	**N**eutrophils	Most common
Mary	**M**onocytes	⇓
Loves	**L**ymphocytes	
Elevating	**E**osinophils	
Beds	**B**asophils	Least common

Also remember that the granulocytes are the "phils" and the agranulocytes are the "cytes," or you could also remember AC-GP.

Fill in the blanks with the correct answers.

FC 25. ____________________ are called leukocytes.

FC 26. The two basic types of lymphocytes are ____________________ and ____________________.

27. ____________________ is an overall decrease in the number of white blood cells (WBCs), where ____________________ is an increase.

28. ____________________ tissue and ____________________ tissue together constitute the hematopoietic, or blood-forming, tissues of the body.

29. ____________________ are really tiny pieces of cells.

FC *Matching – match each leukocyte with the correct description or word (leukocyte may be used more than once).*

	Description
30. ______	Increase with parasitic worms and allergic reactions
31. ______	Immunity
32. ______	Fight primarily against bacteria and viral-infected cells
33. ______	Nuclei may have two, three, or more lobes
34. ______	Produce histamine and heparin
35. ______	Most numerous
36. ______	Largest
37. ______	Cause inflammation and allergies
38. ______	Fight primarily against invading bacteria
39. ______	Produce antibodies

Leukocyte

a. Neutrophils
b. Eosinophils
c. Basophils
d. Lymphocytes
e. Monocytes

Multiple choice – select the best answer.

40. What is it called when leukocytes migrate out of blood vessels and enter tissues?
    a. Diapedesis
    b. Granulocytosis
    c. Leukocyte migration
    d. Leukocytosis

41. Eosinophils are most abundant where?
    a. Liver
    b. Liver and spleen
    c. Lining the respiratory and digestive tracts
    d. Lymph nodes, liver, and spleen

42. Which of the following produce and release histamine and heparin?
    a. Neutrophils
    b. Basophils
    c. Monocytes
    d. Lymphocytes

43. If a person was infected with parasitic worms, which leukocyte would you expect to see increase?
    a. Monocytes
    b. Lymphocytes
    c. Eosinophils
    d. Neutrophils

44. Antibodies come from which leukocyte?
    a. Basophils
    b. Monocytes
    c. Neutrophils
    d. Lymphocytes

45. Which of the following is descriptive of a neutrophil?
    a. Platelet
    b. Phagocyte
    c. Agranulocyte
    d. Erythrocyte

46. Which of the following is the largest leukocyte?
    a. Monocyte
    b. Lymphocyte
    c. Basophil
    d. Neutrophil

47. Aggregation, adhesiveness, and agglutination are properties of:
    a. leukocytes.
    b. granulocytes.
    c. erythrocytes.
    d. thrombocytes.

48. Most of the agranulocytes originate where?
    a. Spleen
    b. Bone marrow
    c. Lymphatic tissue
    d. Liver

49. The blood-forming tissues of the body are called:
    a. myeloid tissue.
    b. lymphatic tissue.
    c. red marrow.
    d. hematopoietic.

FC 50. *Hemostasis* refers to what?
    a. Destroying of blood cells
    b. Creating blood cells
    c. Stoppage of blood flow
    d. Dissolving blood clots

51. What is the first step in hemostasis?
    a. Coagulation
    b. Platelet plug
    c. Agglutination
    d. Thrombopoiesis

52. When platelets encounter _______ in damaged vessel walls, they become sticky and adhere.
    a. collagen
    b. erythrocytes
    c. fibrin
    d. protein

53. What is the formation of platelets called?
    a. Thrombocytosis
    b. Thrombopenia
    c. Thrombopoiesis
    d. Thrombogenesis

54. Thrombocytes originate from:
    a. megakaryoblasts.
    b. myeloblasts.
    c. proerythroblasts.
    d. monoblasts.

Labeling – label the following blood components

55. Composition of whole blood

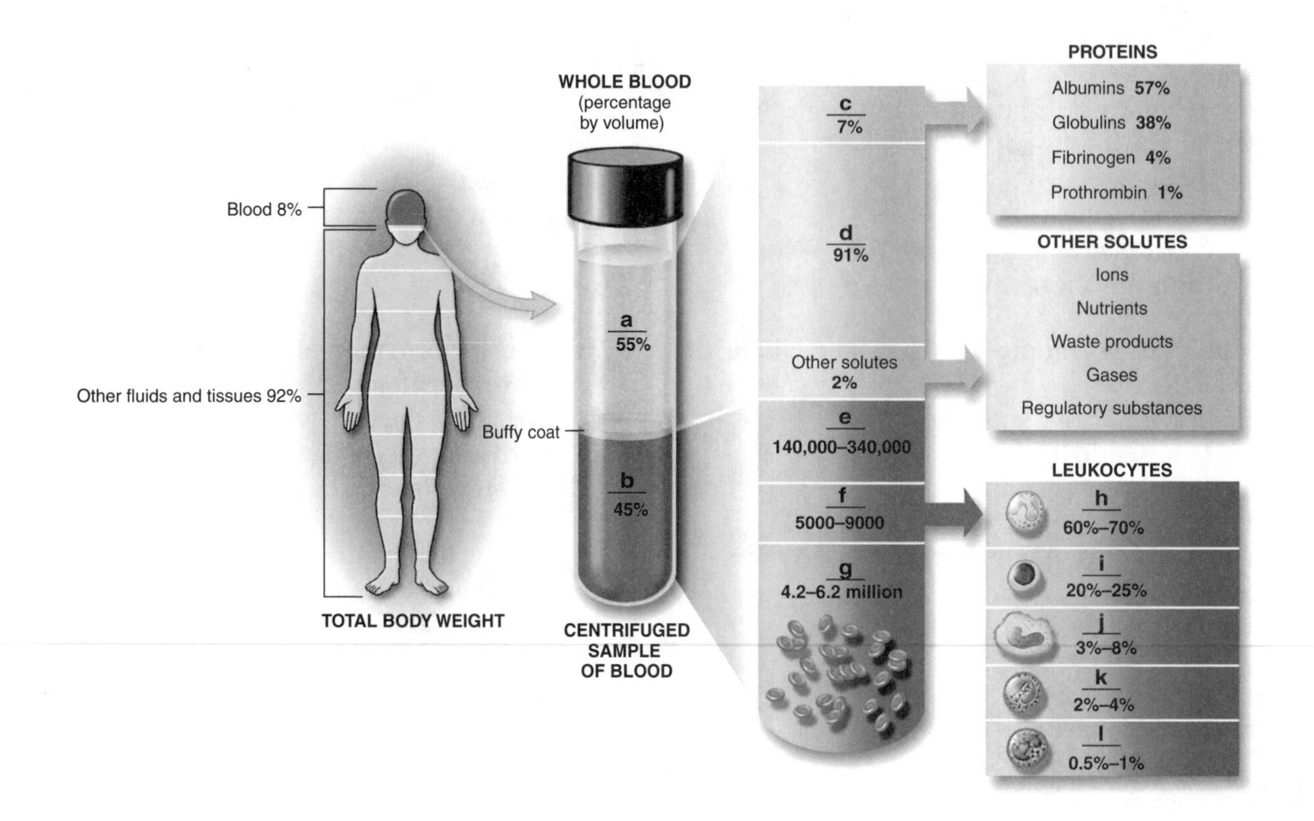

a.	____________	g.	____________
b.	____________	h.	____________
c.	____________	i.	____________
d.	____________	j.	____________
e.	____________	k.	____________
f.	____________	l.	____________

56. The formed elements of blood

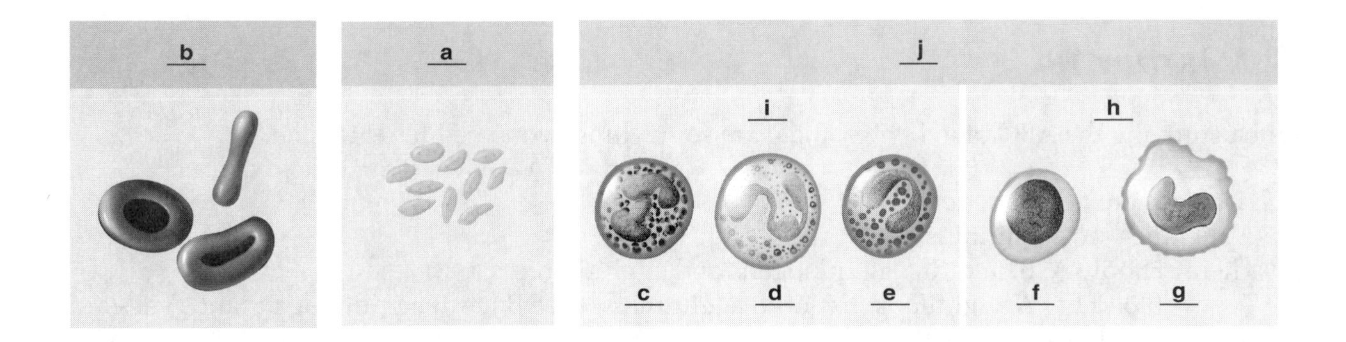

a. ______________________

b. ______________________

c. ______________________

d. ______________________

e. ______________________

f. ______________________

g. ______________________

h. ______________________

i. ______________________

j. ______________________

FC *Short answer*

57. What are the three formed elements of blood?

58. What are the two general types of leukocytes?

59. List the five basic types of leukocytes.

60. List the three granulocytes.

61. List the two agranulocytes.

BLOOD TYPES, BLOOD PLASMA, AND BLOOD CLOTTING

STUDY TIP

When studying the ABO blood types, there are some things you need to remember.

1. Antigens are agglutinogens.
2. Antibodies are agglutinins.
3. There are only A, B, and Rh agglutinogens and agglutinins.
4. Type A blood has A agglutinogens and B agglutinins, type B has B agglutinogens and A agglutinins, and O has zero or 0 (this is how you remember O) agglutinogens and both A and B agglutinins.
5. Rh-positive blood has the Rh agglutinogens and Rh-negative blood does not.
6. When you receive blood, you almost always only receive the cells that do NOT contain the agglutinins. This is called packed red cells (PRC) or a "unit" of blood.
7. You **CANNOT** have the same agglutinogen and agglutinin in your blood or you have a transfusion reaction.
8. Simply **draw** a picture of the blood cell (a circle) and label it according to its blood type. Then label it with the agglutinogens and agglutinins. Whatever is **NOT** on the surface of the cell is the agglutinins and in the plasma. Do this on any questions regarding blood types. See the example below.

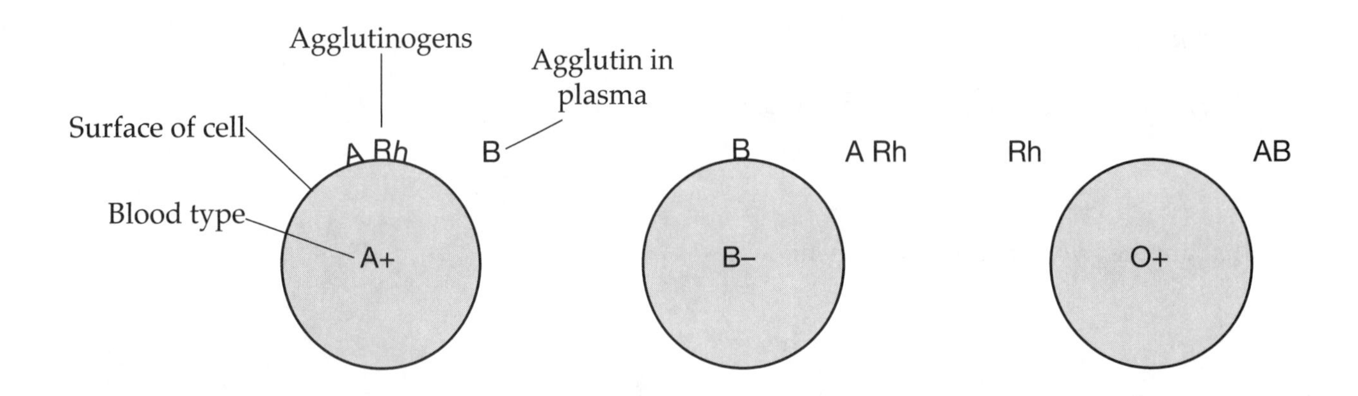

Fill in the blanks with the correct answers.

62. Blood types are determined by the ____________________ present on the plasma membrane.

63. Plasma consists of __________ % water and __________ % solutes and the solutes are classified as ____________________ or ____________________.

64. ____________________ is plasma without its clotting elements.

65. The process of clotting is divided into the ____________________ clotting pathway and ____________________ clotting pathway.

66. ____________________ is synthesized in the intestine by bacteria and is essential to clotting.

FC *Matching – match each term with the correct description.*

	Term		Definition
67. ______	Agglutinin	a.	Dissolves fibrin
68. ______	Agglutinogens	b.	Clot formation
69. ______	Agglutination	c.	RBCs breaking apart
70. ______	Hemolysis	d.	Clumping of RBCs
71. ______	Fibrinolysis	e.	Antibodies
72. ______	Thrombosis	f.	Clot dissolution
73. ______	Plasminogen	g.	Converts plasminogen to plasmin
74. ______	tPA	h.	Found on the plasma membrane
75. ______	Plasmin	i.	Inactive plasma protein
76. ______	Serum	j.	Plasma without the clotting factors

Multiple choice – select the best answer.

77. Which of the following is NOT an antigen in the ABO system?
    a. A
    b. B
    c. O
    d. Rh

78. The term *agglutinin* is often used to describe what?
    a. Antigens
    b. Antibodies
    c. Receptor sites
    d. Transfusion reaction

79. Specific blood antigens are called:
    a. agglutinogens.
    b. agglutinins.
    c. antibodies.
    d. agglutinates.

80. When agglutinins combine and react to agglutinogens it is called:
    a. agglutination.
    b. blood clot.
    c. thrombosis.
    d. hemolysis.

81. What is the "universal donor"?
    a. AB+
    b. AB–
    c. O+
    d. O–

82. What is the "universal recipient"?
    a. AB+
    b. AB–
    c. O+
    d. O–

83. A+ blood has which agglutinogens?
    a. B and Rh
    b. B
    c. A
    d. A and Rh

84. O– blood has which agglutinogens?
    a. A, B, and Rh
    b. A and B
    c. Rh
    d. None

85. Which of the following blood types has no agglutinins?
    a. AB+
    b. AB–
    c. O+
    d. O–

86. A– blood could be given to all the following blood types EXCEPT:
    a. A+.
    b. AB–.
    c. A–.
    d. AB+.

87. Which of the following can receive O– blood?
    a. Only O–
    b. Only O+
    c. All types
    d. O+ and O–

88. Synthesis of most of the plasma proteins occurs where?
    a. Bone marrow
    b. Spleen
    c. Liver
    d. Blood

89. Which of the following is the correct sequence of events in the formation of a clot?
    a. Fibrinogen, fibrin, thrombin, prothrombin activator, prothrombin
    b. Fibrin, fibrinogen, prothrombin, prothrombin activator, thrombin
    c. Prothrombin, prothrombin activator, thrombin, fibrinogen, fibrin
    d. Prothrombin activator, prothrombin, thrombin, fibrin, fibrinogen

90. For the liver to synthesize prothrombin, blood must contain an adequate amount of what?
    a. Vitamin K
    b. Calcium
    c. Thrombin
    d. Prothrombin activator

91. *Fibrinolysis* refers to what?
    a. Formation of fibrin
    b. Dissolving clots
    c. Conversion of fibrinogen to fibrin
    d. Formation of a fibrin clot

92. Formation of a blood clot is called:
    a. thrombosis.
    b. thromboembolus.
    c. fibrinolysis.
    d. thrombocytosis.

93. What hydrolyzes fibrin strands and dissolves blood clots?
    a. Plasmin
    b. tPA
    c. Plasminogen
    d. Fibrinogen

94. What compound impairs the liver's use of vitamin K?
    a. Heparin
    b. tPA
    c. Coumarin
    d. Streptokinase

Short answer

95. List the three main kinds of proteins found in the blood.

96. List the three things that ALL plasma proteins contribute to maintenance of blood.

FC 97. List the four essential components critical to coagulation.

98. List the three substances critical to fibrinolysis.

99. List two anticoagulants.

CHAPTER 17

Anatomy of the Cardiovascular System

HOW TO APPROACH THE ANATOMY OF THE CARDIOVASCULAR SYSTEM

Break this chapter up into the heart and the vessels. The heart is what is going to move the blood through the vessels back to itself. There are two "circuits" or paths of blood flow—the pulmonary and systemic circuits. The same is true of the heart. The right heart is going to the lungs and back (pulmonary) and the left is going to the entire body and back to the right (systemic). When approaching the heart, it helps to remember it is all about the ventricles, especially the left ventricle. It is important and USEFUL to know the circulation of the heart itself (coronaries and their branches). If you have any family with heart disease, you will hear these vessels discussed. The best review is to write on a piece of paper the path of blood from the superior vena cava (SVC) or inferior vena cava (IVC) to the aortic arch and mention every structure on the way. Abbreviate everything that you can to make it quicker and simpler. You can even just make up your own abbreviations for testing purposes. If you can do this, you have a firm grasp of the heart. See the study tip on the valves of the heart.

The blood vessels are mostly about knowing the "correct sequence of blood flow." If you can trace blood from, for example, the tip of your finger to the tip of your toe, you are doing great. Trace the blood through the heart as well and you test both vessels and heart at the same time. Change the path several times to test yourself and as above, use a scratch piece of paper and then look at the pictures to see if you labeled it correctly. Repeat until you miss nothing.

HEART

STUDY TIP

All the organs and the cavities you will learn are covered with membranes. The membrane that lies on the surface of the organ is called *visceral* (meaning organ) and then referenced to the specific organ. Example: the membrane found on the surface of the heart is called *visceral pericardium*, and on the lungs is called *visceral pleura*. The membrane lining the cavity that the organ is in is called *parietal*. Example: parietal pericardium and parietal pleura. The space between them is the cavity: pericardial cavity and pleural cavity.

Fill in the blanks with the correct answers.

FC 1. The heart lies in the ____________________, or middle of the thoracic cavity.

2. The ____________________ is the loose-fitting sac that encloses your heart.

3. ____________________ is produced by the serous pericardium and provides protection against friction as the heart beats.

4. Blood flows into the ____________________ arteries when the ventricles relax.

5. The ____________________ is the pacemaker of the heart.

FC *Matching – match each term with the correct description or word.*

		Term		Definition
6.	______	Sinoatrial (SA) node	a.	Sac around the heart
7.	______	Pericardium	b.	"Pacemaker"
8.	______	Parietal pericardium	c.	Superficial "layer" of the heart
9.	______	Visceral pericardium	d.	"Nonstick" surface
10.	______	Pericardial cavity	e.	Connected to papillary muscles
11.	______	Myocardium	f.	"OUT" valves
12.	______	Atrioventricular (AV) valves	g.	Epicardium
13.	______	Semilunar (SL) valves	h.	Lines the fibrous pericardium
14.	______	Epicardium	i.	Space between parietal and visceral
15.	______	Endocardium	j.	"Heart muscle"

Multiple choice – select the best answer.

16. The apex of the heart points in which direction?
    a. Inferior and medial
    b. Superior and lateral
    c. Inferior and lateral
    d. Superior and medial

17. What membrane lies directly on the heart?
    a. Visceral pericardium
    b. Myocardium
    c. Fibrous pericardium
    d. Parietal layer

18. The layer of the heart that contains intercalated discs is the:
    a. epicardium.
    b. visceral pericardium.
    c. myocardium.
    d. endocardium.

FC 19. The muscular ridges located in the ventricles are called:
    a. trabeculae carneae.
    b. cusps.
    c. chordae tendineae.
    d. pectinate muscles.

20. The venous blood of the heart will empty into the __________ before entering the right atrium.
    a. SVC
    b. IVC
    c. coronary sinus
    d. cardiac vein

FC 21. AV valves connect to papillary muscles via what?
 a. Trabeculae carneae
 b. Chordae tendineae
 c. AV cords
 d. AV tendons

FC 22. Which valves lead into the ventricles?
 a. SL valves
 b. AV valves
 c. Inlet valves
 d. Atrial valves

23. The left ventricular myocardium receives oxygenated blood from the:
 a. left atrium.
 b. SVC.
 c. left coronary artery.
 d. coronary sinus.

24. Contraction of the left atrium causes the blood to open and pass through which valve?
 a. Mitral valve
 b. Tricuspid valve
 c. Aortic SL valve
 d. Pulmonary SL valve

25. Blood in the circumflex artery comes from which of the following vessels?
 a. Left coronary
 b. Anterior interventricular
 c. Ascending aorta
 d. Right coronary

26. Which of the following represents the correct sequence of blood flow through the heart?
 a. Right atrium, mitral valve, right ventricle, aortic SL valve, aorta
 b. Right ventricle, tricuspid valve, right atrium, pulmonary SL valve
 c. Left atrium, mitral valve, left ventricle, pulmonary SL valve, pulmonary trunk
 d. Right atrium, tricuspid valve, right ventricle, pulmonary SL valve, pulmonary trunk

27. Contraction of the left ventricle closes which of the following valves?
 a. Mitral valve
 b. Tricuspid valve
 c. Aortic SL valve
 d. Pulmonary SL valve

STUDY TIP

There are four valves you have to remember. The valves are the **IN** and **OUT** doors to the ventricles. It's all about the ventricles, remember that. The IN valves are called atrioventricular valves or AV valves and they are the tricuspid and mitral. The OUT valves are semilunar (SL) valves and they are the pulmonary and aortic.

Labeling – label the following diagrams.

28. Wall of the heart

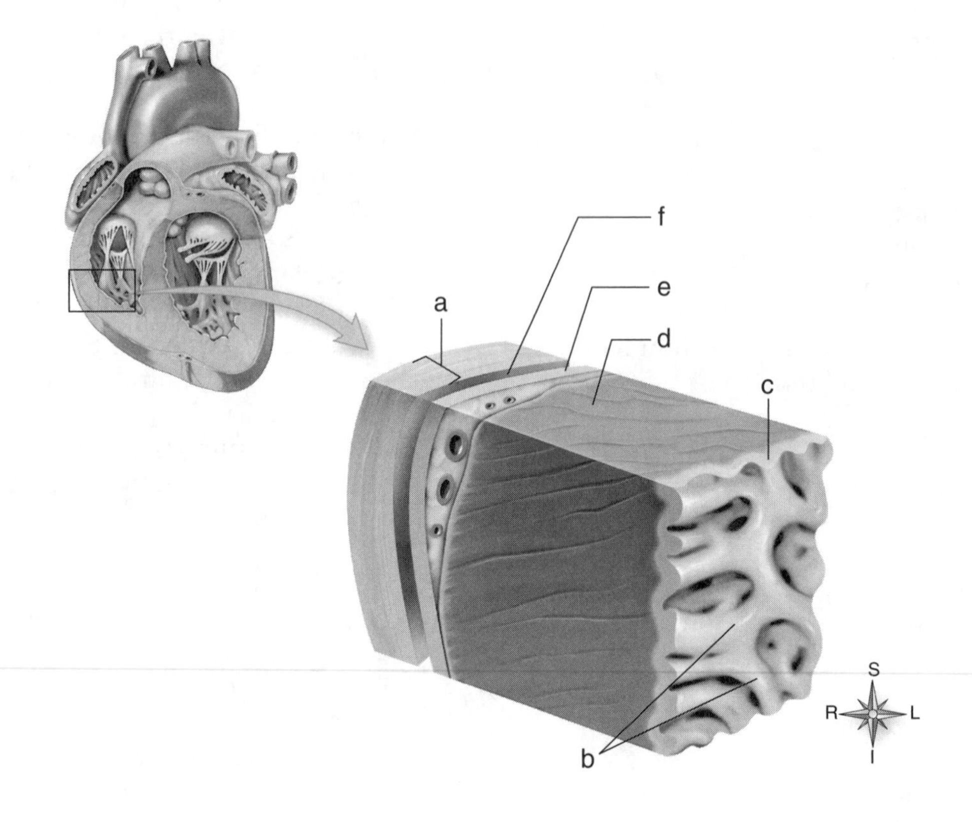

a. ______________________ d. ______________________

b. ______________________ e. ______________________

c. ______________________ f. ______________________

29. The heart and great vessels, anterior view

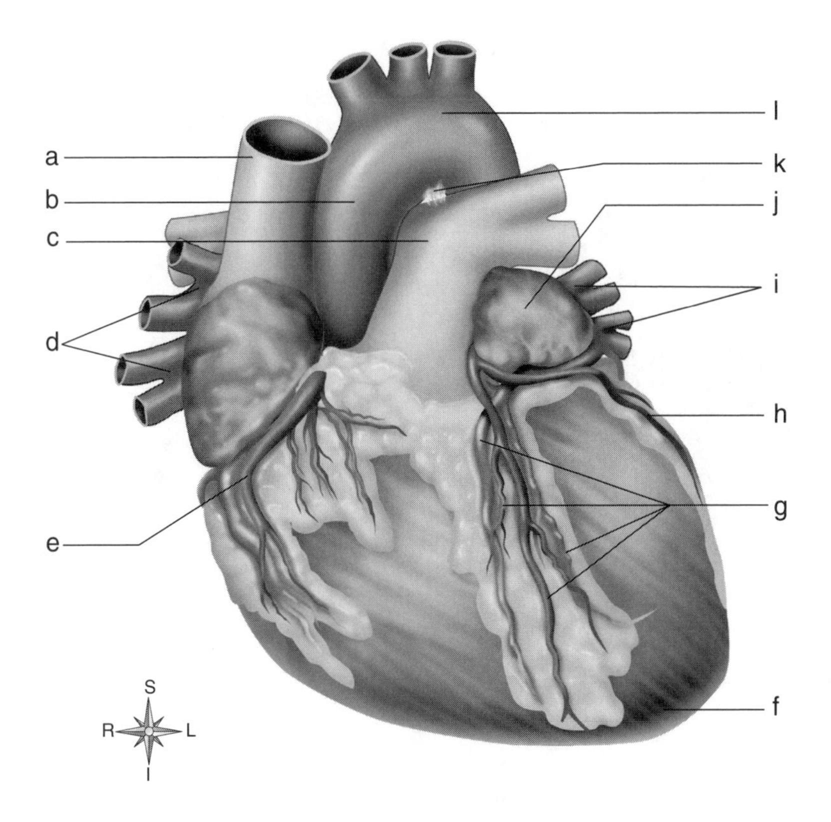

a. ______________________
b. ______________________
c. ______________________
d. ______________________
e. ______________________
f. ______________________
g. ______________________
h. ______________________
i. ______________________
j. ______________________
k. ______________________
l. ______________________

30. Interior of the heart

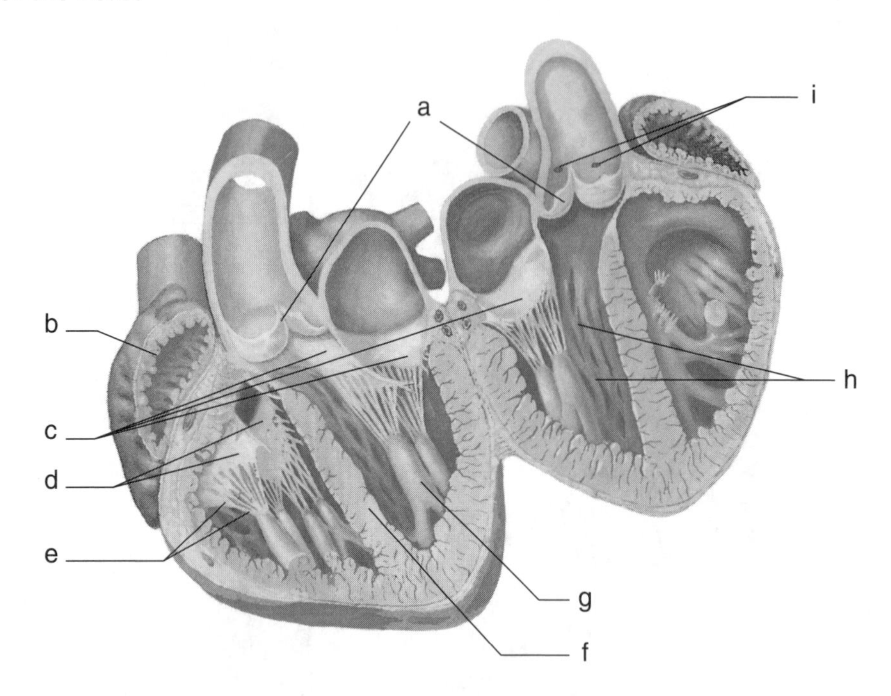

a. ______________________

b. ______________________

c. ______________________

d. ______________________

e. ______________________

f. ______________________

g. ______________________

h. ______________________

i. ______________________

31. Coronary arteries

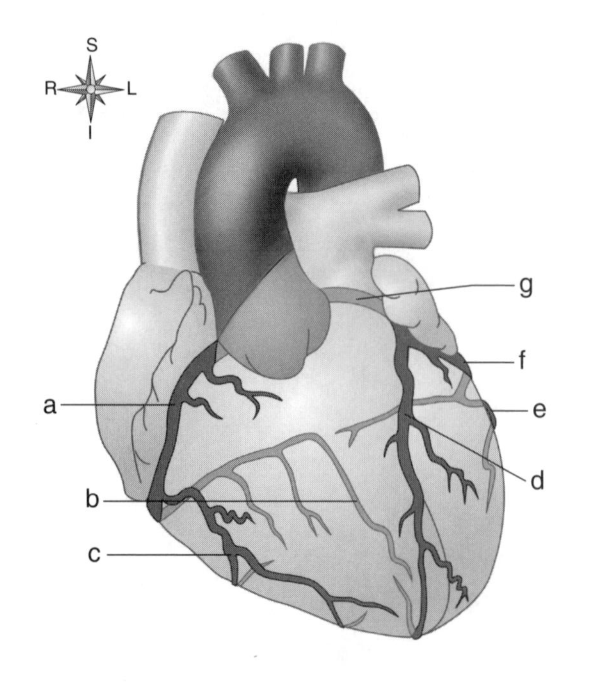

a. ______________________

b. ______________________

c. ______________________

d. ______________________

e. ______________________

f. ______________________

g. ______________________

Short answer

32. Name the two coverings of the heart.

FC 33. List the three layers of the heart wall.

FC 34. List the four chambers of the heart.

FC 35. Name the two AV valves.

FC 36. Name the two SL valves.

BLOOD VESSELS

Fill in the blanks with the correct answers.

FC 37. ____________________ take blood AWAY from the heart and ____________________ take blood TO the heart and ____________________ connect the two.

38. ____________________ act as valves regulating blood flow into specific capillaries.

39. ____________________ provide structure and strength but very limited elasticity in vessel walls.

40. Smooth muscle is found in the walls of the entire vascular system except in ____________________.

41. The brachial, femoral, and gastric arteries are examples of ____________________ arteries.

HINT

STUDY TIP

Remember – **A**rteries – **A**way, and Veins – To. They are not classified by level of O_2 in the blood, but direction of flow. Both have the same three layers starting with tunica: externa (external), media (middle), and intima (internal). Also **V**eins have **V**alves and arteries do not.

Matching – match each term with the correct description.

	Term		Definition
42. ______	True capillaries	a.	Largest arteries
43. ______	Arteries	b.	"To"
44. ______	Arterioles	c.	Receives blood from metarterioles
45. ______	Venules	d.	Smallest arteries
46. ______	Elastic arteries	e.	"Connecting vessels"
47. ______	Muscular arteries	f.	Tiny veins
48. ______	Metarterioles	g.	Smallest blood vessels
49. ______	Capillaries	h.	No blood vessels
50. ______	Veins	i.	"Away"
51. ______	Avascular	j.	Most of the smaller arteries

Multiple choice – select the best answer.

52. The aorta and some other large arteries are what type of artery?
    a. Muscular arteries
    b. Elastic arteries
    c. Arterioles
    d. Metarterioles

53. These are connecting vessels that connect true arterioles with capillaries.
    a. True capillaries
    b. Sinusoids
    c. Metarterioles
    d. Precapillary arterioles

54. Which of the following are considered the "primary exchange vessels"?
    a. Arteries
    b. Veins
    c. Capillaries
    d. Venules

55. Which of the following receive blood directly from the metarterioles?
    a. True capillaries
    b. Arterioles
    c. Sinusoids
    d. Venules

56. What type of vessel is called *capacitance vessels*?
    a. Veins
    b. Arteries
    c. Capillaries
    d. Lymphatic

57. Which of the following vessels serve as a reservoir for blood and can contain as much as 60% of total blood volume?
    a. Veins
    b. Arteries
    c. Capillaries
    d. Lymphatic

58. Which type of tissue provides a smooth, nonstick surface?
    a. Collagen fibers
    b. Endothelial cells
    c. Elastic fibers
    d. Smooth muscle cells

59. Smooth muscle is primarily in which layer of blood vessels?
    a. Tunica externa
    b. Tunica endothelium
    c. Tunica media
    d. Tunica intima

Labeling – label the following diagram.

60. Structure of blood vessels

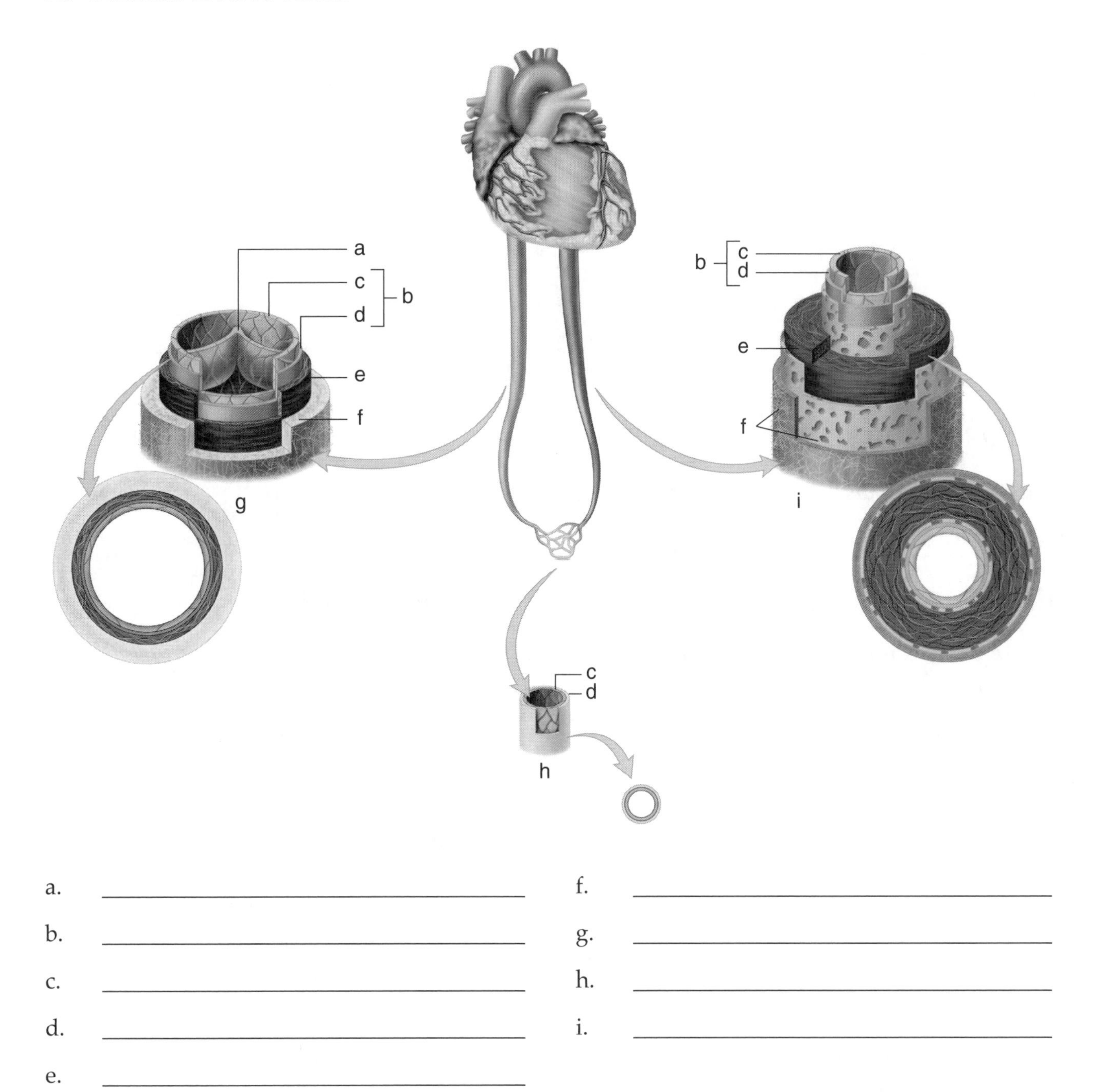

a. ______________________

b. ______________________

c. ______________________

d. ______________________

e. ______________________

f. ______________________

g. ______________________

h. ______________________

i. ______________________

Short answer

FC 61. List the three types of vessels.

62. Name four types of arteries.

63. Name the four types of tissue "fabrics" that make up a typical vessel wall.

FC 64. List the three layers of blood vessels profundus to superficial.

MAJOR BLOOD VESSELS - ARTERIES

Fill in the blanks with the correct answers.

FC 65. ____________________ conducts blood through all parts of the body.

FC 66. ____________________ conducts blood just from the heart to the lungs and back.

67. A direct connection or merger of blood vessels to one another is called a ____________________.

FC 68. The ____________________ is the major artery that serves as the main trunk for the entire systemic arterial system.

Matching – match each "region supplied" with the correct artery.

		Arteries		Region supplied
69.	______	Coronary	a.	Lower arm and medial hand
70.	______	Superficial and deep palmar arches	b.	Lower arm and lateral hand
71.	______	Common carotid	c.	Hand and fingers
72.	______	Brachial	d.	Abdominal viscera
73.	______	Radial	e.	Myocardium
74.	______	Ulnar	f.	Descending colon, rectum
75.	______	Celiac trunk	g.	Head and neck
76.	______	Inferior mesenteric	h.	Pelvis
77.	______	Common iliac	i.	Arm and hand
78.	______	Internal iliac	j.	Pelvis and lower extremity

Multiple choice – select the best answer.

79. Blood passing through two consecutive capillary beds is called a:
    a. portal system.
    b. bisystemic circulation.
    c. bicapillary system.
    d. vascular anastomosis.

FC 80. Which of the following are the three branches of the celiac trunk?
    a. Superior mesenteric, inferior mesenteric, and suprarenal
    b. Gastric, splenic, hepatic
    c. Renal, ovarian, testicular
    d. Inferior phrenic, lumbar, median sacral

FC 81. Which of the following supplies blood to the ear and the hair on your head?
    a. Vertebral
    b. External carotid
    c. External iliac
    d. Subclavian

82. Complete the sequence: subclavian, __________, brachial, radial, palmar arches.
    a. ulnar
    b. deep brachial
    c. vertebral
    d. axillary

83. Which of the following vessels leads to middle rectal, vaginal, superior gluteal, and internal pudendal?
    a. External iliac
    b. Popliteal
    c. Lateral sacral
    d. Internal iliac

84. Blood in the anterior tibial artery just came from which artery?
    a. Posterior tibial
    b. Popliteal
    c. Plantar arch
    d. Femoral

85. Which of the following is the correct flow of blood?
    a. Brachiocephalic, right subclavian, axillary, brachial, ulnar
    b. Brachiocephalic, left subclavian, axillary, brachial, ulnar
    c. Brachiocephalic, right internal carotid, external carotid, vertebral
    d. Brachiocephalic, left common carotid, internal carotid, external carotid

86. Which of the following is the correct flow of blood?
    a. Abdominal aorta, celiac trunk, external iliac, femoral, popliteal
    b. Abdominal aorta, common iliac, femoral, popliteal, anterior tibial
    c. Abdominal aorta, celiac trunk, superior mesenteric, inferior mesenteric
    d. Abdominal aorta, common iliac, internal iliac, femoral, posterior tibial

87. What two arteries "feed" the brain?
    a. Vertebral and internal carotid
    b. External carotid and internal carotid
    c. Brachiocephalic and common carotid
    d. Vertebral and common carotid

Labeling – label the following diagrams.

88. Major arteries of the head and neck

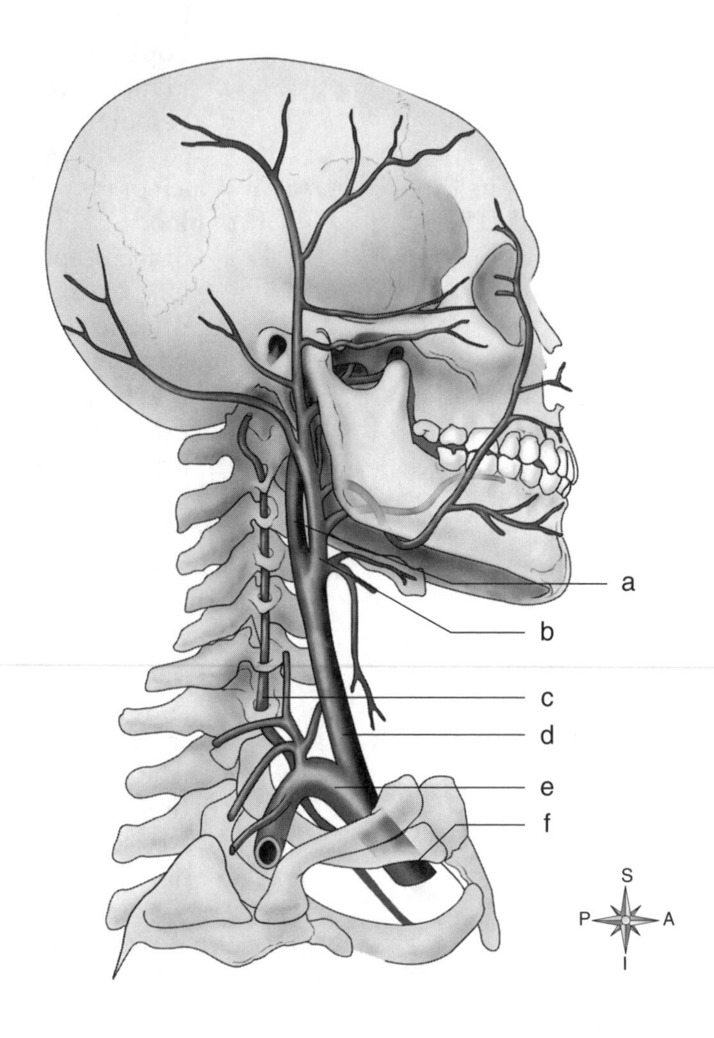

a. ____________________

b. ____________________

c. ____________________

d. ____________________

e. ____________________

f. ____________________

89. Divisions and primary branches of the aorta

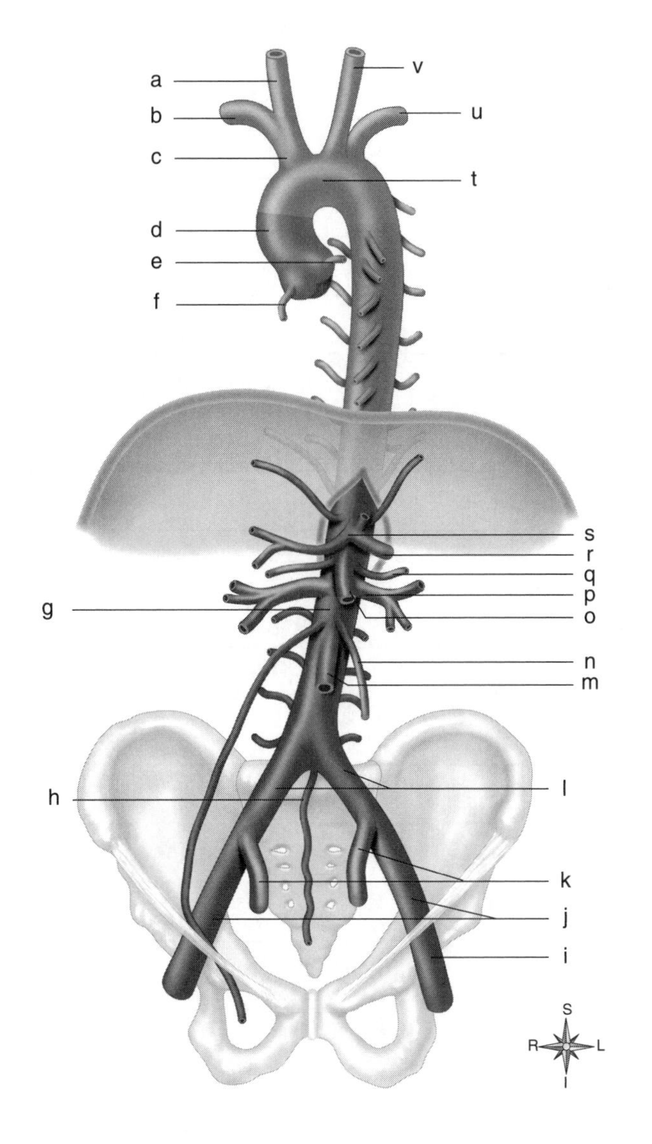

a. ______________________
b. ______________________
c. ______________________
d. ______________________
e. ______________________
f. ______________________
g. ______________________
h. ______________________
i. ______________________
j. ______________________
k. ______________________
l. ______________________
m. ______________________
n. ______________________
o. ______________________
p. ______________________
q. ______________________
r. ______________________
s. ______________________
t. ______________________
u. ______________________
v. ______________________

90. Principal arteries of the body

q
p
a
b
c
d
e
o
n
m
l
k
j
f
g
h
S
R
L
I
i

a. ______________________________

b. ______________________________

c. ______________________________

d. ______________________________

e. ______________________________

f. ______________________________

g. ______________________________

h. ______________________________

i. ______________________________

j. ______________________________

k. ______________________________

l. ______________________________

m. ______________________________

n. ______________________________

o. ______________________________

p. ______________________________

q. ______________________________

FC *Short answer*

91. Name five arteries located on the arm.

92. Name four arteries located on or going to the neck or head.

93. Name five of the visceral branches of the abdominal aorta.

94. List six arteries of the leg.

95. Name all the sections (5) of the aorta.

VEINS AND FETAL CIRCULATION

Fill in the blanks with the correct answers.

96. The ____________________ arteries are temporary extensions of the internal iliac arteries and carry fetal blood to the placenta.

97. The ____________________ is a continuation of the umbilical vein.

98. The ____________________ drains the upper body back to the heart and the ____________________ drain the lower body back to the heart.

99. The ____________________ receives blood from the mediastinum and pericardium.

Matching – match region drained with the correct vein.

	Vein		Region
100. ______	Internal jugular	a.	Lateral and lower arm and hand
101. ______	External jugular	b.	Liver
102. ______	Cephalic	c.	Descending colon and rectum
103. ______	Basilic	d.	Brain
104. ______	Hepatic	e.	Superficial medial and anterior thigh, leg, and foot
105. ______	Inferior mesenteric	f.	Superficial head and neck
106. ______	Superior mesenteric	g.	Small intestine and most of colon
107. ______	Common iliac	h.	Pelvis
108. ______	Great saphenous	i.	Lower extremities and pelvis
109. ______	Internal iliac	j.	Medial and lower arm and hand

Multiple choice – select the best answer.

110. As respiration is established after birth, what becomes the fibrous cord, the ligamentum ateriosum?
    a. Ductus venosus
    b. Ductus arteriosus
    c. Foramen ovale
    d. Fossa ovalis

111. What connects the cephalic to the basilic vein?
    a. Brachial
    b. Palmar venous arches and median cubital
    c. Palmar venous arches
    d. Median cubital

FC 112. Which of the following is an opening in the septum between the right and left atria?
a. Ductus arteriosus
b. Fossa ovalis
c. Foramen ovale
d. Ductus venosus

FC 113. Which of the following connects the pulmonary artery to the aorta?
a. Umbilical vein
b. Ductus arteriosus
c. Foramen ovale
d. Ductus venosus

FC 114. Which of following is found in the adult atrial septum?
a. Fossa ovalis
b. Ligamentum arteriosum
c. Foramen ovale
d. Ligamentum venosum

FC 115. What is the main vein that drains the brain?
a. Facial
b. External jugular
c. Cephalic
d. Internal jugular

116. Which of the following veins lead INTO the liver?
a. Hepatic
b. Cephalic
c. Superior mesenteric
d. Hepatic portal

FC 117. The most medial vein of the leg is the:
a. great saphenous.
b. small saphenous.
c. fibular.
d. posterior tibial.

118. The most lateral vein of the arm is the:
a. radial.
b. ulnar.
c. basilic.
d. cephalic.

119. Which of the following veins brings blood back from the diaphragm?
a. Phrenic
b. Cystic
c. Pulmonary
d. Azygos

120. What vein is formed by the joining of the brachial and basilic?
a. Subclavian
b. Cephalic
c. Axillary
d. Median cubital

121. The great saphenous merges with which of the following veins?
a. Femoral
b. External iliac
c. Internal iliac
d. Common iliac

122. What two veins merge with the splenic vein?
a. Superior and inferior mesenteric
b. Inferior mesenteric and pancreatic
c. Gastric and pancreatic
d. Cystic and gastroepiploic

Labeling – label the following diagrams.

123. Inferior vena cava and its abdominopelvic tributaries

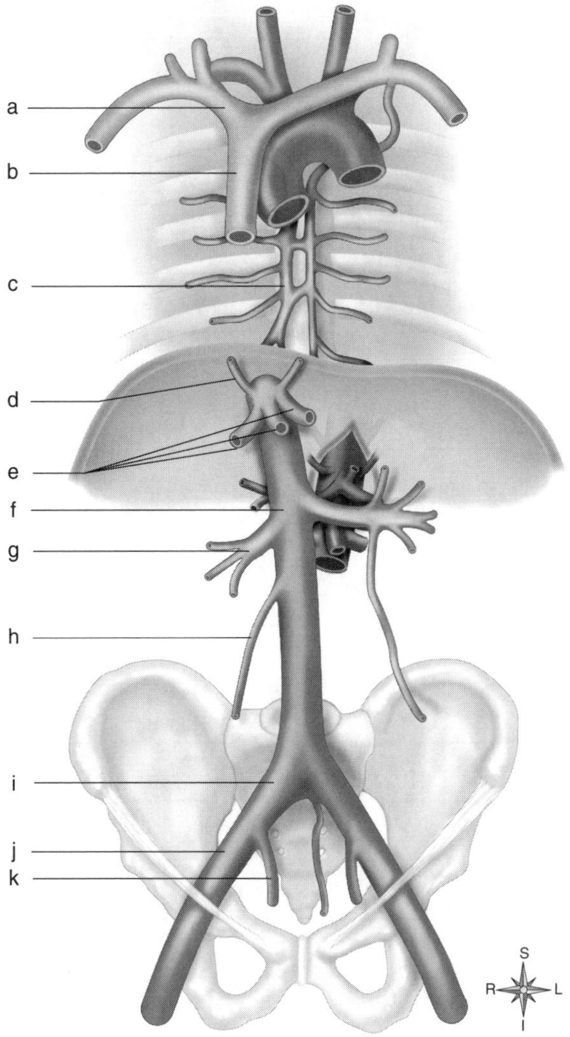

a. ______________________________

b. ______________________________

c. ______________________________

d. ______________________________

e. ______________________________

f. ______________________________

g. ______________________________

h. ______________________________

i. ______________________________

j. ______________________________

k. ______________________________

124. Principal veins of the body

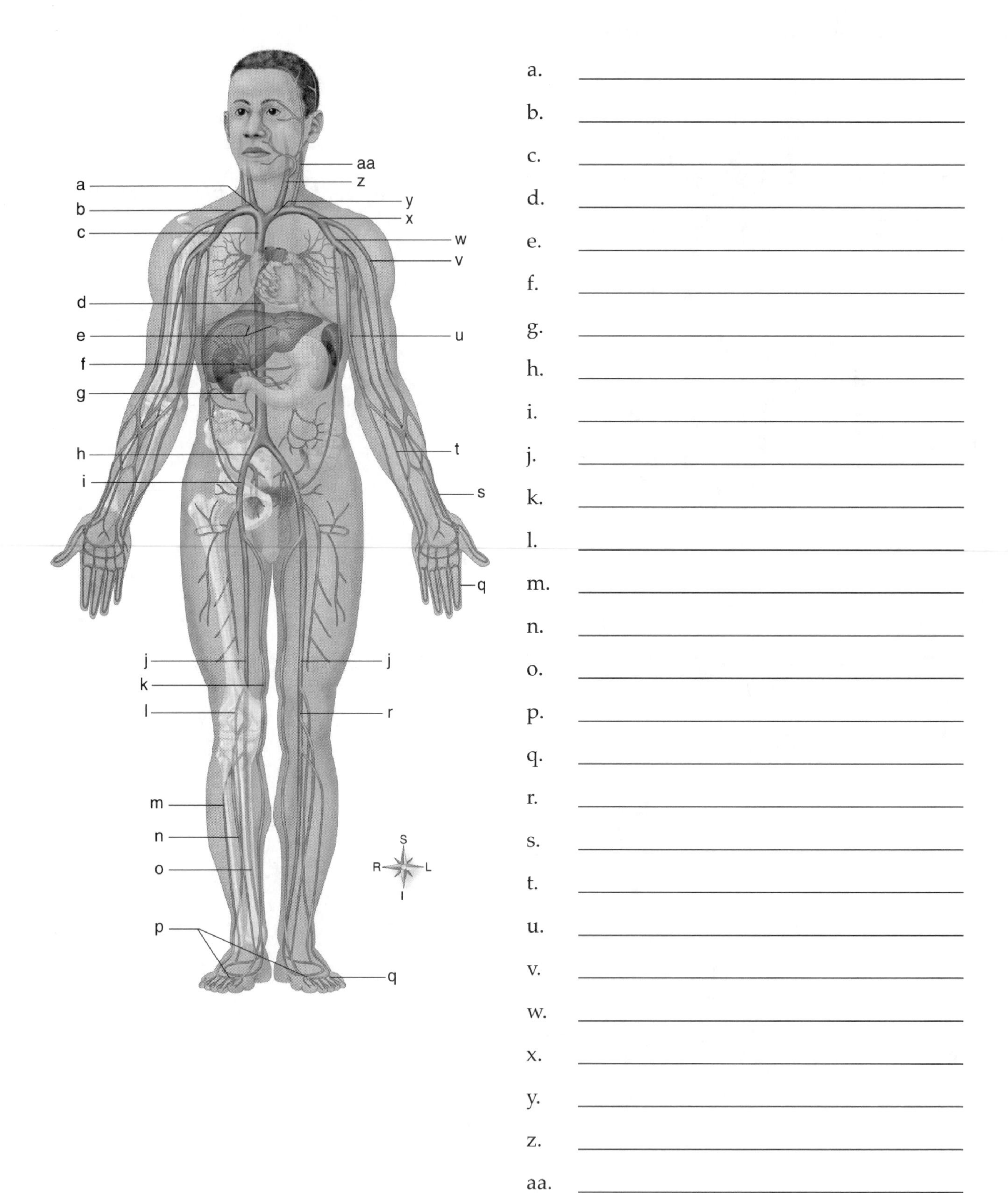

a. ______________________________

b. ______________________________

c. ______________________________

d. ______________________________

e. ______________________________

f. ______________________________

g. ______________________________

h. ______________________________

i. ______________________________

j. ______________________________

k. ______________________________

l. ______________________________

m. ______________________________

n. ______________________________

o. ______________________________

p. ______________________________

q. ______________________________

r. ______________________________

s. ______________________________

t. ______________________________

u. ______________________________

v. ______________________________

w. ______________________________

x. ______________________________

y. ______________________________

z. ______________________________

aa. ______________________________

125. Plan of fetal circulation

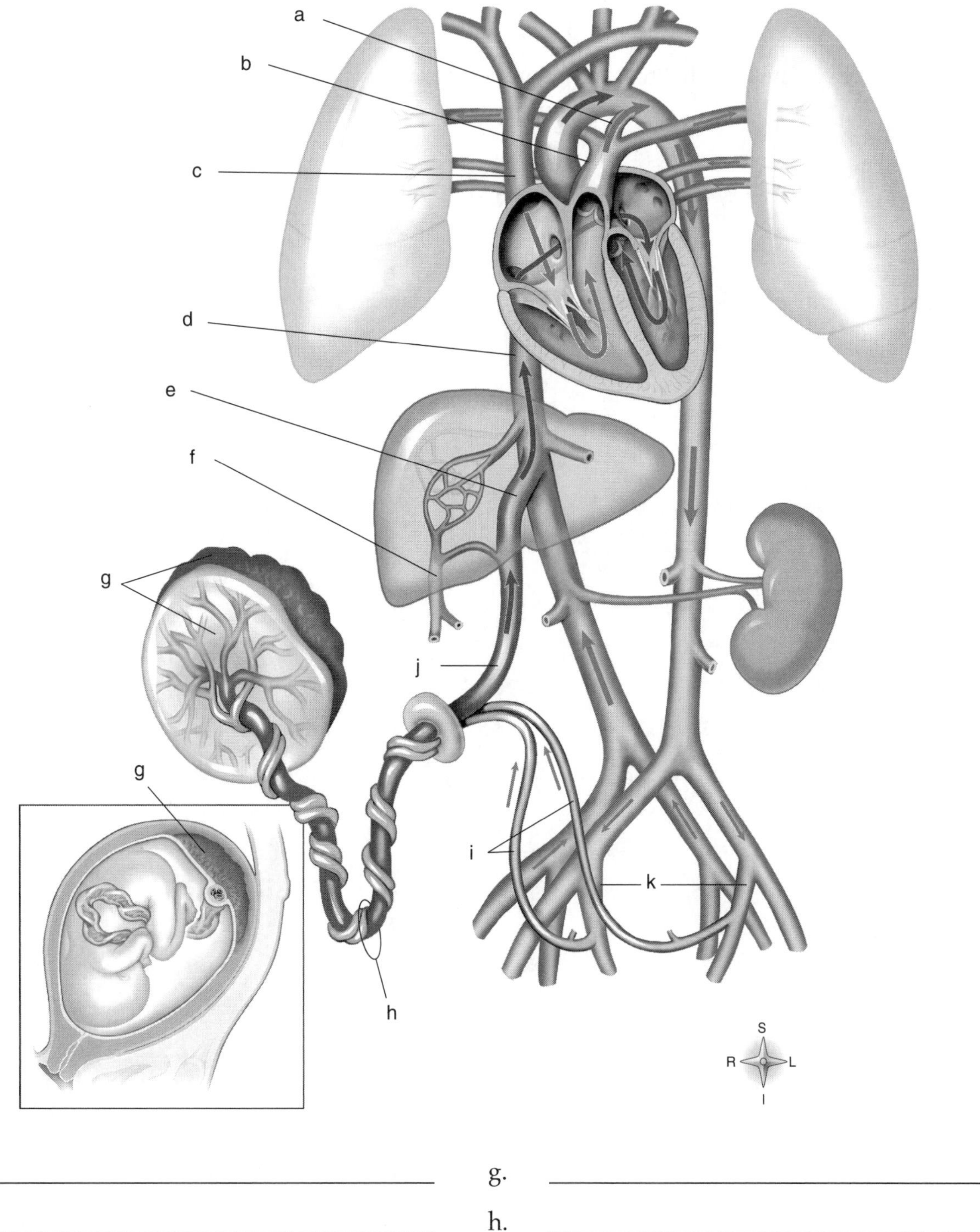

a. ______________________

b. ______________________

c. ______________________

d. ______________________

e. ______________________

f. ______________________

g. ______________________

h. ______________________

i. ______________________

j. ______________________

k. ______________________

Short answer

FC 126. Name five veins of the arm.

127. Name the seven veins that form the hepatic portal vein.

128. Name five veins that drain into the internal jugular.

129. Name the three veins that form the transverse sinus.

CHAPTER 18

Physiology of the Cardiovascular System

HOW TO APPROACH THE PHYSIOLOGY OF THE CARDIOVASCULAR SYSTEM

The main focus of this chapter is the electrical conduction system of the heart. Be sure to know the path of the impulse as it travels through the heart and what is happening at each point. There are lots of abbreviations used in this chapter that you need to know. The electrocardiogram (ECG) is a very useful tool in the medical profession and is frequently used, so be sure to understand the basics of what it is measuring and what it can tell you. *Systole* and *diastole* are terms that are important to understanding both the heart and blood pressure (BP). Make sure you understand what BP represents, how to measure it, what makes it go up and down, and what the body does to regulate and monitor it.

HEMODYNAMICS AND THE HEART AS A PUMP

Fill in the blanks with the correct answers.

FC 1. ____________________ refers to the various mechanisms that influence the movement of blood.

2. The ____________________ conduct the impulses through the muscle of the ventricles.

FC 3. The impulse conduction of the heart is measured on a(n) ____________________.

4. The ____________________ describes a complete heartbeat or single pumping cycle.

5. ____________________ is the period between closing of the semilunar (SL) valves and the opening of the ____________________ valves.

6. A(n) ____________________ is simply an "abnormal" heart sound that can be caused by many things.

FC *Matching – match the correct description, word, or phrase with each term.*

	Term		Definition
7. ______	Residual volume	a.	"Relaxation"
8. ______	Sinoatrial (SA) node	b.	Contraction of the ventricles
9. ______	Atrioventricular (AV) node	c.	"Contraction"
10. ______	AV bundle	d.	Slow the impulse
11. ______	Subendocardial branches	e.	Purkinje fibers
12. ______	"Lubb"	f.	"Pacemaker"
13. ______	"Dubb"	g.	Conducts impulse to subendocardial branches
14. ______	Systole	h.	Blood remaining in ventricles after contraction
15. ______	Diastole	i.	Closing of the AV valves
16. ______	QRS complex	j.	Closing of the SL valves

Multiple choice – select the best answer.

17. The interatrial bundle does what?
    a. Slows the impulse
    b. Increases the conduction velocity as it is relayed through the AV bundle
    c. Transmits the impulse through the ventricles
    d. Facilitates rapid conduction to the left atrium

18. Which of the following slows the impulse, allowing the atria to contract completely before reaching the ventricles?
    a. SA node
    b. AV node
    c. AV bundle
    d. Subendocardial branches

19. Pacemakers other than the SA node are called:
    a. secondary pacemakers.
    b. ectopic pacemakers.
    c. dysrhythmic pacemakers.
    d. adjunct pacemakers.

20. An ECG is measuring:
    a. pulse rate.
    b. heart contractions.
    c. movement of the heart.
    d. electrical events of the heart.

21. Which of the following represents depolarization of the atria?
    a. P wave
    b. ST segment
    c. QRS
    d. T wave

22. The T wave represents:
    a. depolarization of the atria.
    b. repolarization of the atria.
    c. depolarization of the ventricles.
    d. repolarization of the ventricles.

23. Contraction of the ventricles is called:
    a. QRS.
    b. systole.
    c. diastole.
    d. stroke volume.

24. Atrial systole is represented by the:
    a. P wave.
    b. QRS.
    c. T wave.
    d. PR interval.

25. The first heart sound, "lubb," is:
    a. an audible contraction of the ventricles.
    b. closing of the AV valves.
    c. an audible contraction of the atria.
    d. closing of the SL valves.

26. Which of the following forces blood to open the AV valves?
   a. Atrial systole
   b. Atrial diastole
   c. Ventricular systole
   d. Ventricular diastole

FC 27. Which of the following is the correct impulse pathway through the heart?
   a. AV node, SA node, Purkinje fibers, bundle branches, AV bundle
   b. SA node, AV bundle, AV node, bundle branches, Purkinje fibers
   c. SA node, AV node, AV bundle, bundle branches, Purkinje fibers
   d. AV node, SA node, bundle branches, AV bundle, Purkinje fibers

Labeling – label the following diagrams.

28. Conduction system of the heart

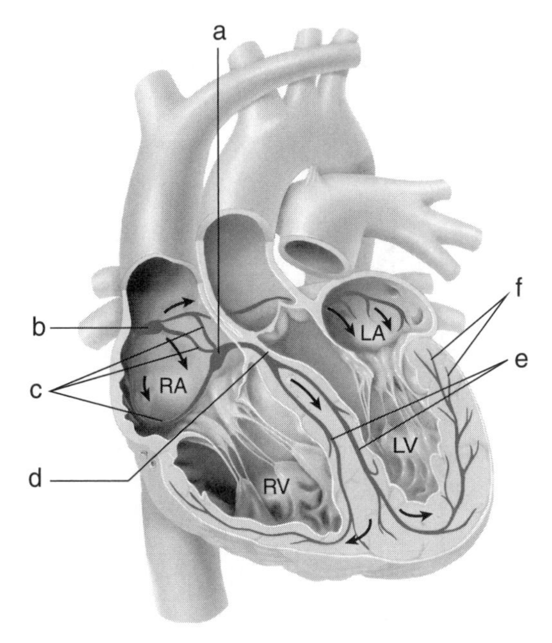

a. ______________________
b. ______________________
c. ______________________
d. ______________________
e. ______________________
f. ______________________

29. Electrocardiogram deflection waves

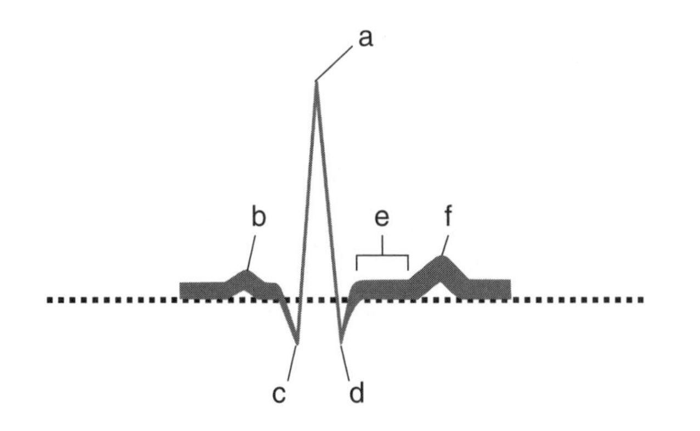

a. ______________________
b. ______________________
c. ______________________
d. ______________________
e. ______________________
f. ______________________

PRIMARY PRINCIPLES OF CIRCULATION

Fill in the blanks with the correct answers.

30. Two of the most important factors in determining BP other than volume are ____________________ and ____________________.

FC 31. Cardiac output is determined by ____________________ and ____________________.

FC 32. ____________________ stimulation slows heart rate (HR) and decreases contractility.

33. Blood ____________________ increases with more "solids" such as cells and proteins in the blood.

34. Reducing the vessel diameter is called ____________________, whereas increasing vessel diameter is called ____________________.

35. A(n) ____________________ in pressure in the aortic and carotid baroreceptors increases impulses to the medulla, causing ____________________.

Matching – match each term with the correct description.

	Term		Definition
36. ______	Pulse pressure	a.	Expansion throughout the arteries
37. ______	Pulse wave	b.	Ventricular relaxation
38. ______	Diastolic BP	c.	SV x HR
39. ______	Baroreceptors	d.	Increasing vessel diameter
40. ______	Vasodilation	e.	Blood pumped out of a ventricle each beat
41. ______	Systolic BP	f.	Ventricular contraction
42. ______	Atrial natriuretic hormone (ANH)	g.	"Force" of contraction
43. ______	Contractility	h.	Difference between systolic and diastolic pressures
44. ______	Stroke volume (SV)	i.	Released due to increased pressures
45. ______	Cardiac output	j.	Sensitive to changes in pressure

STUDY TIP

HR usually does the opposite of BP. If your BP suddenly drops, your HR increases (like when you get off the couch too fast and get lightheaded and feel your heart pounding). If your BP suddenly increases, your HR decreases.

Multiple choice – select the best answer.

46. Increased contractility of the heart causes which of the following?
    a. Decreased cardiac output (CO)
    b. Increased SV
    c. Increased residual volume
    d. Increased resistance

FC 47. Which of the following is the correct method of calculating CO?
    a. SV x HR
    b. SV / HR
    c. SV + HR
    d. SV – HR

FC 48. The amount of blood pumped by the heart per minute is called:
    a. HR.
    b. SV.
    c. CO.
    d. ejection fraction.

FC 49. Frank-Starling law states:
    a. The faster the heart beats, the greater the CO.
    b. The more the heart is stretched, the greater the contraction.
    c. The less the heart is stretched, the greater the contraction.
    d. The more the heart is stretched, the lesser the contraction.

FC 50. The strength of the heart contraction is called:
    a. Frank-Starling law.
    b. contractility.
    c. chronotropic.
    d. cardiac force.

51. Which of the following will decrease HR?
    a. Anxiety
    b. Fear
    c. Anger
    d. Decreased blood temperature

52. Which of the following will increase HR?
    a. Vagal stimulation
    b. Decrease in BP
    c. Decrease in blood temperature
    d. Sudden, intense pain

FC 53. The vasomotor center that regulates BP is located in the:
    a. heart.
    b. vessel walls.
    c. medulla.
    d. carotids.

54. A sudden increase in arterial BP causes:
    a. HR to decrease.
    b. HR to increase.
    c. vasoconstriction.
    d. decrease in contractility.

55. What is the main factor that influences venous return to the heart?
    a. Respirations
    b. Gravity
    c. Muscle contraction
    d. Viscosity

FC 56. What causes release of ANH?
    a. Overstretching of the atrial wall
    b. Increased contractility
    c. Increased BP
    d. Decreased HR

57. A sphygmomanometer measures what?
    a. HR
    b. BP
    c. Contractility
    d. Electrical activity

58. What is the first sound heard when measuring blood pressure?
    a. Diastolic BP
    b. Arterial BP
    c. HR
    d. Systolic BP

59. What is the pulse pressure if the BP is 120/80?
    a. 200
    b. 60
    c. 40
    d. 1.5

60. Which of the following is defined as alternating expansion and recoil of an artery?
    a. BP
    b. Pulse pressure
    c. Pulse
    d. Pulse wave

Short answer

FC 61. Name three factors critical to arterial BP.

FC 62. What are the two baroreceptors important to maintaining BP?

FC 63. Name five important factors that influence HR.

64. Name two factors that influence peripheral resistance.

FC 65. List three things that influence venous return to the heart.

66. Name six palpable pulses.

STUDY TIP

BP and HR are fairly simple to understand. Just think of things that happen to you and how your body reacts. Example, what happens to your BP and HR when you get scared, tired, get up from bed too quickly, get dehydrated, etc.? If you think about these examples, you most likely will be able to answer the majority of the questions related to this topic.

CHAPTER 19

Lymphatic and Immune Systems

HOW TO APPROACH THE LYMPHATIC AND IMMUNE SYSTEMS

The lymphatic system is designed to gather fluid lost from circulation and direct it to lymph nodes to be cleaned by the immune system. Think of lymph nodes as filters cleaning interstitial fluid as it makes its way back to general circulation. This is why your lymph nodes get large when the part of the body that drains to them is infected. When the lymphatic system does not work, the fluid cannot get back to general circulation and we swell up (edema). The lymphatic system is almost identical to the venous system in its structure.

The immune system is your second line of defense and is divided into nonspecific and specific. Think about "nonspecific" and "specific" and it makes sense. To understand the immune system you MUST have a solid foundation of the basic terms such as *antigens, antibodies, phagocytosis, antigen presentation,* etc. When you understand the basics, the rest comes pretty easily. Many of the terms explain themselves such as "naturally acquired" immunity or "artificially acquired" immunity. When you get confused or lost, go back to the basics that you understand and then move forward again. This really is fun and the chapter is applicable to life. Example: can you really get the "flu" from the flu shot? At the end of this chapter, you should be able to put this myth to rest and explain to others why. Remember for the specific immune system to work, the pathogen must be found and phagocytized in order to present the antigen to B and T cells. It's all about the antigens! No antigens, no immunity.

LYMPHATIC SYSTEM

Fill in the blanks with the correct answers.

FC 1. The ____________________ prevents fluid retention by acting like a drainage system for ____________________ fluid.

2. The walls of lymphatic capillaries consist of a single layer of ____________________ cells.

3. The milky lymph containing high concentrations of fat found in lacteals after digestion is called ____________________.

4. The center of a lymph node is composed of ____________________ that increase surface area and are lined with macrophages.

FC 5. ____________________ are the sites of both biological and mechanical filtration.

6. ____________________ of the spleen contains nodules of developing lymphocytes and the ____________________ is comprised of blood-filled sinuses.

Matching – match each description, word, or phrase with the correct term.

		Term		Definition
FC 7.	_____	Lymph	a.	Defend our bodies from pathogens and provide sites for maturation of lymphocytes
FC 8.	_____	Interstitial fluid	b.	Paired
9.	_____	Chyle	c.	"Adenoids"
10.	_____	Lymph nodes	d.	Base of the tongue
11.	_____	Palatine tonsils	e.	Responsible for maturing T cells
12.	_____	Pharyngeal tonsils	f.	Most active in very young children
13.	_____	Lingual tonsils	g.	Recycles erythrocytes and platelets
FC 14.	_____	Thymus	h.	Lymph containing lots of fat from the intestine
FC 15.	_____	Thymosin	i.	Fills spaces between cells
FC 16.	_____	Spleen	j.	Originated from interstitial fluid

Multiple choice – select the best answer.

17. Lymph nodes associated with the small intestine, especially the ileum, are called:
    a. aggregated lymphoid nodules.
    b. lacteals.
    c. intestinal lacteals.
    d. peritoneal patches.

18. Lymphatic vessels that originate in the intestine are called:
    a. lymphatic capillaries.
    b. lacteals.
    c. lymphatic arteries.
    d. intestinal lymphatic capillaries.

19. The thoracic duct drains lymph from the entire:
    a. body except the right upper quarter.
    b. upper body.
    c. lower body.
    d. left side of the body.

20. The right lymphatic duct receives lymph from the right:
    a. side of the body.
    b. side of the head and neck.
    c. upper quadrant.
    d. upper thorax and arm.

21. The lymphatic ducts drain lymph back into circulation through which of the following vessels?
    a. Aorta
    b. Subclavian veins
    c. Superior vena cava
    d. Brachiocephalic veins

22. Which of the following that accumulates in the tissue space can only return to the blood by the lymphatic system?
    a. Ions
    b. Glucose
    c. Proteins
    d. Red blood cells (RBCs)

FC 23. Lymph is formed from fluid from where?
    a. Lungs
    b. Thoracic duct
    c. Interstitial fluid
    d. Veins

24. Which of the following is NOT a function of lymph?
    a. Defense
    b. Return tissue fluid to blood
    c. Transport oxygen back to blood
    d. Intestinal absorption of fats and fat-soluble vitamins

25. Where might a person have an infection if their superficial cubital lymph node is enlarged?
    a. Throat
    b. Toe
    c. Finger
    d. Breast (mastitis)

FC 26. Impeding the flow of lymph causes:
    a. elevated blood pressure.
    b. decreased intracellular fluid.
    c. edema.
    d. increased intravascular fluid.

27. One of the greatest concerns with breast cancer is:
    a. immunosuppression.
    b. spreading to the axillary lymph nodes.
    c. painful surgery.
    d. cancer spreading to the head and face.

FC 28. When you look in your mouth, what tonsils are you able to see?
    a. Lingual
    b. Pharyngeal
    c. Adenoids
    d. Palatine

29. This structure is located near the opening of the nasal cavity in the upper throat.
    a. Pharyngeal tonsil
    b. Lingual tonsil
    c. Thymus gland
    d. Palatine tonsil

FC 30. What structure is concerned with the processing and maturation of T lymphocytes?
    a. Tonsils
    b. Lymph nodes
    c. Thymus gland
    d. Spleen

31. Which of the following is located just below the diaphragm and just above the left kidney?
    a. Liver
    b. Thymus gland
    c. Spleen
    d. Thoracic duct

32. Which of the following is lined with macrophages that remove microorganisms and worn-out RBCs?
    a. Tonsils
    b. Thymus gland
    c. Spleen
    d. Lymph nodes

Labeling – label the following diagrams.

33. Principal organs of the lymphatic system

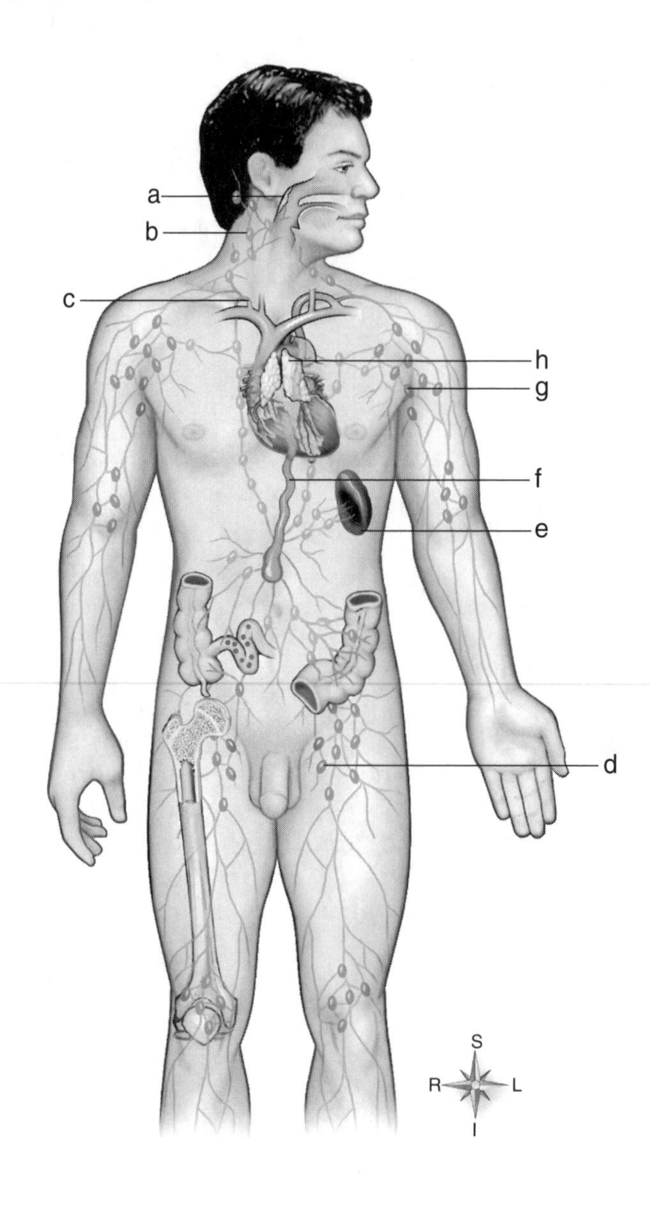

a. ______________________

b. ______________________

c. ______________________

d. ______________________

e. ______________________

f. ______________________

g. ______________________

h. ______________________

34. Circulation plan of lymphatic fluid

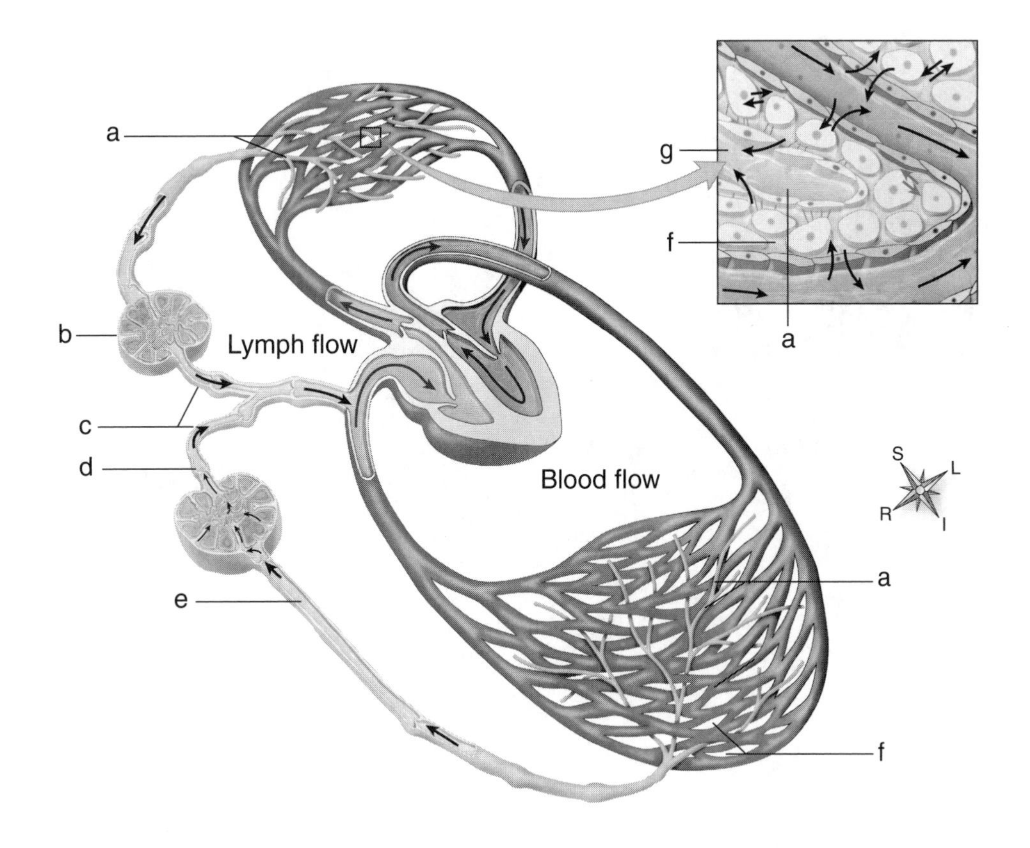

a. ______________________

b. ______________________

c. ______________________

d. ______________________

e. ______________________

f. ______________________

g. ______________________

35. Location of the tonsils

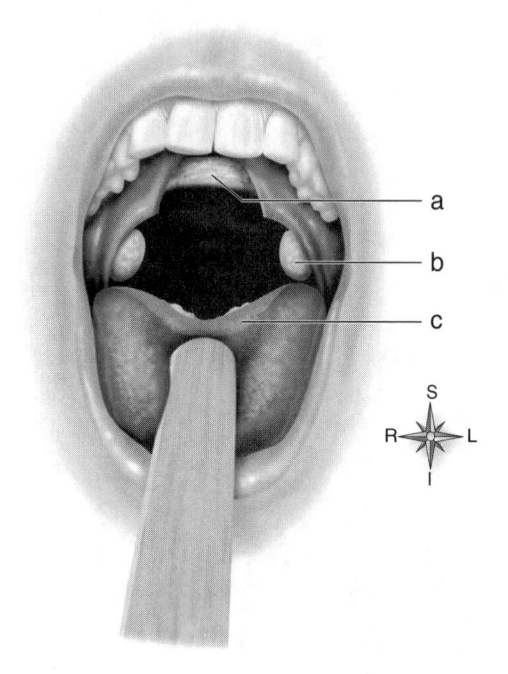

a. ______________________

b. ______________________

c. ______________________

FC *Short answer*

36. List three functions of the lymphatic system.

37. What three things aid lymph in flowing toward the heart?

38. List four functions of the spleen.

39. List the three tonsils.

40. Name five major lymphoid organs or structures.

IMMUNE SYSTEM

Fill in the blanks with the correct answers.

FC 41. Molecules found on the surfaces of cells that can be used to identify them are called ______________________.

42. The ability of our immune system to ignore our own body cells is called ______________________.

FC 43. ____________________ is the ingestion and destruction of microorganisms.

44. ____________________ molecules trigger the fever response by promoting production of certain prostaglandins.

FC *Matching – match each description, word, or phrase with the correct term.*

	Term		Definition
45. ______	Nonspecific immunity	a.	"Alarm" to nearby cells to protect them from viruses
46. ______	Specific immunity	b.	Large monocyte that has migrated out of the bloodstream
47. ______	Complement proteins	c.	Mechanical and chemical barriers, inflammation, fever, phagocytes, and NK cells
48. ______	Natural killer (NK) cells	d.	Cells that digest invading organisms and display their proteins to activate other immune cells
49. ______	Antigen-presenting cells (APCs)	e.	Very effective against tumor cells and cells infected by viruses
50. ______	Phagocytosis	f.	Ingestion and destruction of microbes
51. ______	Interferon	g.	Attracting other leukocytes
52. ______	Chemotaxis	h.	Lyses cells and promotes opsonization
53. ______	Cytokines	i.	Chemicals responsible for chemotaxis
54. ______	Macrophage	j.	T and B cells

Multiple choice – select the best answer.

55. Chemicals released from cells to trigger or regulate an immune response are called:
    a. immunokines.
    b. cytokines.
    c. chemosomes.
    d. chemokines.

56. All of the following are examples of mechanical/chemical barriers EXCEPT:
    a. skin.
    b. mucus.
    c. hair.
    d. sebum.

57. The process of attracting leukocytes through the use of chemicals is called:
    a. inflammatory migration.
    b. leukomigration.
    c. chemotaxis.
    d. leukotaxis.

58. A pyrogen is a molecule that causes what?
    a. Fever
    b. Inflammation
    c. Illness
    d. Chemotaxis

59. The process of marking microbes with complement proteins for phagocytic cells to destroy is called:
    a. opsonization.
    b. diapedesis.
    c. adaptive immunity.
    d. antibody-mediated immunity.

60. A phagosome is a:
    a. pathogen that has been encapsulated by pseudopods.
    b. cell capable of phagocytosis of pathogens.
    c. footlike projection that is an extension of the cell membrane.
    d. stem cell that is capable of becoming a phagocyte.

FC 61. The movement of phagocytes from the blood to the tissue is called:
    a. leukomigration.
    b. diapedesis.
    c. pavementing.
    d. phagopedesis.

62. Cells that process and display the proteins of pathogens on their cell membranes are called:
    a. APCs.
    b. NK cells.
    c. phagosomes.
    d. interferons.

FC 63. Which of the following cells are critical in preventing cancer?
    a. Macrophages
    b. Phagocytes
    c. NK cells
    d. Neutrophils

FC 64. Cells invaded by viruses will synthesize a protein that protects uninfected cells. This glycoprotein is called:
    a. complement protein.
    b. interferon.
    c. Major histocompatibility complex (MHC).
    d. virogen.

FC 65. Which of the following proteins lyses foreign cells?
    a. Interferons
    b. MHCs
    c. Phagocytes
    d. Complement

Short answer

66. List three cell types involved in innate immunity.

67. Give three examples of cytokines.

68. List five mediators of inflammation.

FC 69. List five examples of nonspecific immunity.

ADAPTIVE IMMUNITY

Fill in the blanks with the correct answers.

FC 70. The two major types of lymphocytes involved in adaptive immunity are ____________________ and ____________________.

FC 71. ____________________ attack pathogens or direct other cells to attack them and are activated by ____________________ on the surface of the pathogens.

FC 72. Antibodies are actually proteins called ____________________.

73. The binding of an antigen to an antibody creates a(n) ____________________.

74. Presentation of an antigen by an APC activates the T cell to form ____________________ T cells and ____________________ T cells.

Matching – match each description, word, or phrase with the correct term.

	Term		Definition
FC 75. ______	T lymphocytes	a.	Antibody-mediated immunity
FC 76. ______	Plasma cells	b.	Will produce plasma cells when exposed to antigens again
77. ______	Memory B cells	c.	Secrete massive amounts of antibodies
FC 78. ______	B lymphocytes	d.	Series of proteins activated by antibodies
FC 79. ______	Complement	e.	Present antigens to T and B cells
FC 80. ______	Active immunity	f.	Developed in another individual
FC 81. ______	Passive immunity	g.	Own immune system responds
82. ______	IgE	h.	Immunoglobulin
83. ______	APCs	i.	Cell-mediated immunity

STUDY TIP

The way to remember the immunoglobulins is "GAMED." Not sure what it means, but a student came up with it and it is rather hard to forget.

Multiple choice – select the best answer.

84. The *CD system* refers to what?
    a. A specific type of antibodies
    b. Cell differentiation
    c. Specific lymphocytes that produce antibodies against pathogens
    d. A way of classifying lymphocytes by surface proteins

85. Which of the following is NOT a location where you might expect to find a lot of lymphocytes?
    a. Adrenal gland
    b. Spleen
    c. Thymus gland
    d. Bone marrow

FC 86. Which of the following B cells produce antibodies?
    a. Killer B cells
    b. Memory B cells
    c. Plasma cells
    d. CD cells

87. How many types of B lymphocytes are there?
    a. 1
    b. 2
    c. 3
    d. 4

88. How many types of T lymphocytes are there?
    a. 1
    b. 2
    c. 3
    d. 4

89. Immunoglobulins are shaped like a:
    a. V.
    b. W.
    c. Y.
    d. T.

90. Which of the following is NOT an effect of antibodies?
    a. Activation of complement proteins
    b. Agglutination
    c. Stimulating phagocytosis
    d. Producing immunoglobulins

91. Which of the following is the most abundant of the immunoglobulins and is involved with "immunizations"?
    a. IgM
    b. IgE
    c. IgG
    d. IgD

92. Which of the following immunoglobulins is involved with allergies?
    a. IgE
    b. IgM
    c. IgA
    d. IgG

FC 93. How many classes of immunoglobulins are there?
    a. 3
    b. 4
    c. 5
    d. 6

FC 94. When organisms stick together in clumps due to antibodies, this is called:
    a. agglutination.
    b. complement cascade.
    c. globulin cascade.
    d. an epitope.

FC 95. Which of the following is responsible for forming membrane attack complexes (MACs)?
    a. Immunoglobulins
    b. Complement proteins
    c. Antibodies
    d. T-lymphocytes

96. Complement proteins can form a "donut" with a hole in it, killing the pathogen. This is called:
    a. complement attack molecule.
    b. membrane attack complexes (MACs).
    c. antibody attack complex (AACs).
    d. epitope.

FC 97. Lymphocytes that travel to the thymus gland become:
    a. T cells.
    b. plasma cells.
    c. B cells.
    d. post T cells.

98. Which of the following is true?
    a. T cells can only react to protein fragments presented on the surface of APCs.
    b. T cells phagocytize and present the antigen to antibodies.
    c. B cells present antigens to T cells to initiate the immune response.
    d. B cells, when presented with an antigen, will produce complement proteins for that antigen.

99. Perforin does what?
    a. Produces a ringlike hole in the plasma membrane of target cells
    b. Causes apoptosis
    c. Attracts other T cells and B cells
    d. Causes antibodies to bind with antigens

100. Getting chickenpox as a child will give you which of the following types of immunity?
    a. Natural passive
    b. Natural active
    c. Artificial passive
    d. Artificial active

101. Immunity that military personnel receive immediately prior to being deployed in the form of a shot is which type of immunity?
    a. Natural passive
    b. Natural active
    c. Artificial passive
    d. Artificial active

FC *Short answer*

102. List the five classes of immunoglobulins.

103. List three effects of antibodies.

104. List the four types of immunity.

CHAPTER 20

Respiratory System

HOW TO APPROACH THE RESPIRATORY SYSTEM

The respiratory system has one main purpose: to exchange gases (breathing). Everything is designed to accomplish this task. The upper respiratory tract is designed to clean, warm, and humidify the air. The lower respiratory tract is for exchanging the gases. You will need to memorize the path that the air takes from mouth or nose to the alveoli in the lungs. If you can trace this path, you most likely have 90% of the anatomy down. The next thing to understand is the physiology of HOW we breathe. You will need to be sure to master the basic concepts of positive and negative pressures. Once you understand the pressures, the mechanics of breathing are relatively easy to understand. Lastly, make sure you have a firm grasp on how the gases move. Go back and review diffusion if you are not 100% confident on this concept. If you are good on diffusion, you should not have any difficulty on movement of gases. See the study tip below on partial pressures.

UPPER RESPIRATORY TRACT

Fill in the blanks with the correct answers.

1. Each nasal cavity is divided into three passageways called _____________________.

2. The nostrils open into an area called the _____________________.

3. The air-containing spaces that open into the nasal cavity are called _____________________.

FC 4. The _____________________ lies between the base of the tongue and the upper end of the trachea.

FC *Matching – match each description, word, or phrase with the correct term.*

	Term		Definition
5. ______	Upper respiratory tract	a.	"Voicebox"
6. ______	Lower respiratory tract	b.	False vocal cords
7. ______	Turbinates	c.	Respiratory and digestive
8. ______	Paranasal sinuses	d.	Larynx, pharynx, nose
9. ______	Pharynx	e.	Opening between vocal folds
10. ______	Larynx	f.	Trachea, bronchi, lungs
11. ______	Vestibular folds	g.	Increase surface area
12. ______	Vocal folds	h.	Blocks trachea when swallowing
13. ______	Glottis	i.	Four pairs
14. ______	Epiglottis	j.	True vocal cords

Multiple choice – select the best answer.

15. Which of the following is NOT part of the upper respiratory tract?
    a. Larynx
    b. Pharynx
    c. Trachea
    d. Nose

FC 16. What serves as a common pathway for the respiratory and digestive tracts?
    a. Pharynx
    b. Larynx
    c. Esophagus
    d. Laryngopharynx

17. What extends from the hyoid bone to the esophagus?
    a. Larynx
    b. Pharynx
    c. Oropharynx
    d. Laryngopharynx

18. All of the following are part of the nasal septum EXCEPT:
    a. maxillary bone.
    b. ethmoid bone.
    c. vomer bone.
    d. septal nasal and vomeronasal cartilages.

19. The meatuses or passageways in the nasal cavity are created by:
    a. projection of the conchae from the medial walls.
    b. projection of the conchae from the lateral walls.
    c. opening of the sinuses from the medial walls.
    d. opening of the sinuses from the lateral walls.

20. Which of the following is the correct flow of air?
    a. Anterior nares, vestibule, meatuses, posterior nares, pharynx
    b. Nostrils, meatuses, vestibule, anterior nares, posterior nares, pharynx
    c. Vestibule, anterior nares, meatuses, posterior nares, pharynx
    d. Nostrils, meatuses, anterior nares, posterior nares, vestibule, pharynx

21. Which mucous membrane is near the roof of the nasal cavity and is paler and has a yellowish tint due to fewer blood vessels?
    a. Respiratory mucosa
    b. Paranasal mucosa
    c. Olfactory epithelium
    d. Paranasal epithelium

22. Which of the following is NOT a function of the nose?
    a. Filtered
    b. Dried
    c. Warmed
    d. Chemically examined

23. All of the following are divisions of the pharynx EXCEPT:
    a. laryngopharynx.
    b. nasopharynx.
    c. oropharynx.
    d. sinopharynx.

FC 24. The space between the vocal folds is the:
    a. glottis.
    b. false vocal folds.
    c. true vocal folds.
    d. vocal cords.

25. How many total cartilages form the framework of the larynx?
    a. 6
    b. 8
    c. 9
    d. 11

26. What is the most anterior cartilage of the larynx?
    a. Thyroid
    b. Epiglottis
    c. Arytenoid
    d. Cricoid

FC 27. What is the largest cartilage of the larynx?
    a. Thyroid
    b. Epiglottis
    c. Arytenoid
    d. Cricoid

28. Which of the following is NOT a function of the larynx?
    a. Filter air
    b. Voice production
    c. Entrance for solids and liquids
    d. Moisten air

Labeling – label the following diagrams.

29. Upper respiratory tract

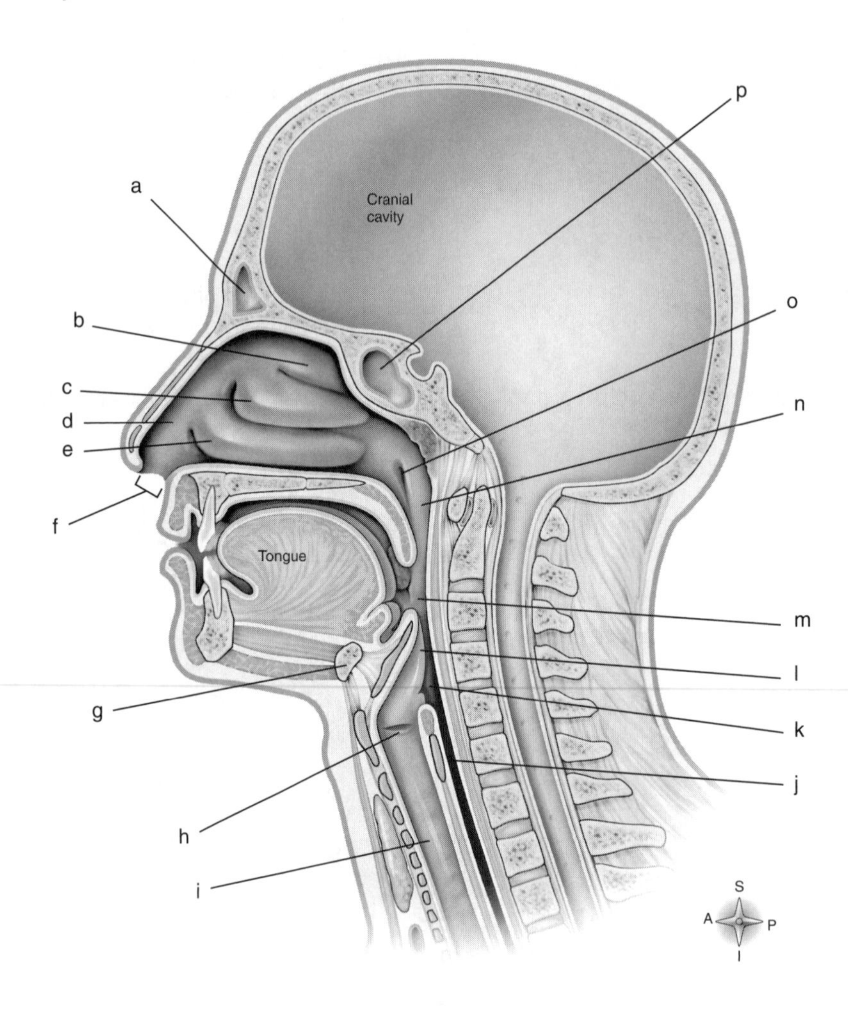

a.	______________________	i.	______________________
b.	______________________	j.	______________________
c.	______________________	k.	______________________
d.	______________________	l.	______________________
e.	______________________	m.	______________________
f.	______________________	n.	______________________
g.	______________________	o.	______________________
h.	______________________	p.	______________________

30. Larynx

a. ______________________

b. ______________________

c. ______________________

d. ______________________

e. ______________________

f. ______________________

g. ______________________

h. ______________________

i. ______________________

j. ______________________

k. ______________________

31. Laryngeal cartilages

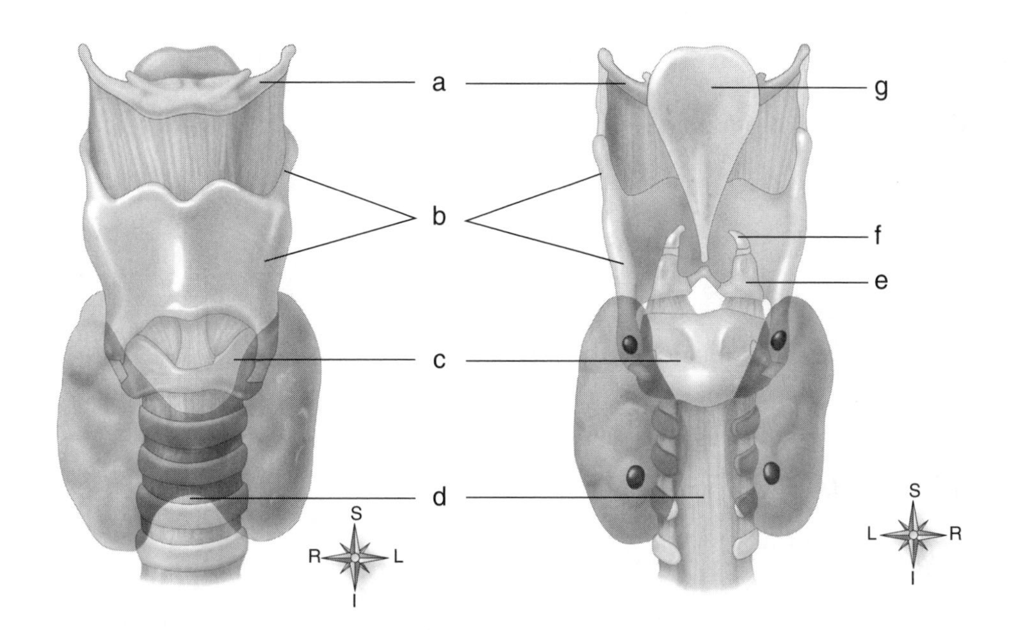

a. ______________________

b. ______________________

c. ______________________

d. ______________________

e. ______________________

f. ______________________

g. ______________________

32. Vocal folds

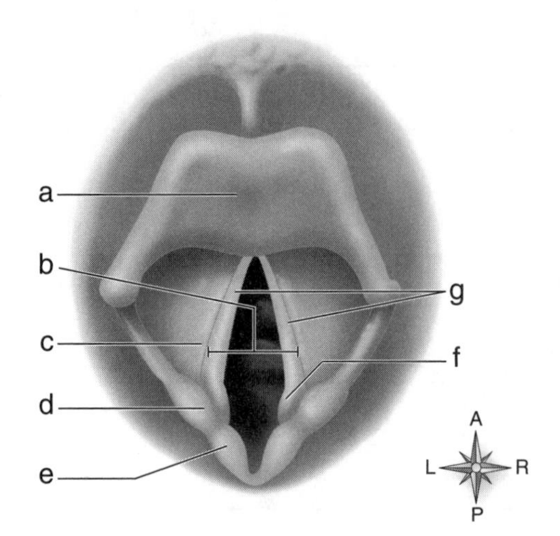

a. ____________________

b. ____________________

c. ____________________

d. ____________________

e. ____________________

f. ____________________

g. ____________________

FC *Short answer*

33. List seven functions of the respiratory system.

34. Name five major structures of the upper respiratory tract.

35. Name three major structures of the lower respiratory tract.

36. List the four pairs of paranasal sinuses.

37. Name the three divisions of the pharynx.

38. List the six different cartilages of the larynx.

LOWER RESPIRATORY TRACT

Fill in the blanks with the correct answers.

39. The ____________________ provides a sturdy open passageway from the upper respiratory tract into the lungs.

FC 40. The force of attraction between water molecules is called ____________________.

41. The epithelium of the trachea is ____________________.

42. The trachea divides into two primary bronchi, the ____________________ bronchus being slightly larger and more vertical than the ____________________.

Matching – match each description, word, or phrase with the correct term.

	Term		Definition
43. ______	Parietal pleura	a.	Where everything enters the lung
44. ______	Visceral pleura	b.	"Inferior" part of lung
45. ______	Bronchopulmonary segment	c.	On the surface of lung
46. ______	Apex	d.	Lining lung cavity
47. ______	Base	e.	"Tube" entering lung
48. ______	Hilum	f.	Functional unit of lung
49. ______	Surfactant	g.	"Air sacs" making up the lung
50. ______	Alveoli	h.	"Superior" part of lung
51. ______	Primary bronchi	i.	Reduces surface tension
52. ______	Trachea	j.	"Windpipe"

Multiple choice – select the best answer.

53. The location where everything enters the lung is called:
    a. hilum.
    b. base.
    c. apex.
    d. pleural cavity.

54. Which of the following is made up of a single layer of simple squamous epithelial tissue?
    a. Terminal bronchioles
    b. Trachea
    c. Alveoli
    d. Bronchioles

55. Which of the following is the correct flow of air into the lungs?
    a. Trachea, primary bronchi, secondary bronchi, bronchioles, tertiary bronchi
    b. Tertiary bronchi, bronchioles, terminal bronchioles, respiratory bronchioles, alveolar sacs
    c. Tertiary bronchi, secondary bronchi, primary bronchi, bronchioles, terminal bronchioles
    d. Secondary bronchi, tertiary bronchi, terminal bronchioles, respiratory bronchioles, bronchioles

FC 56. The inner surface of the alveoli is coated with a fluid containing:
    a. lipids.
    b. surfactant.
    c. mucus.
    d. immune enzymes.

57. Which of the following lines the entire thoracic cavity?
    a. Parietal pleura
    b. Visceral pleura
    c. Pulmonary pleura
    d. Thoracic pleura

Labeling – label the following diagrams.

58. Respiratory system

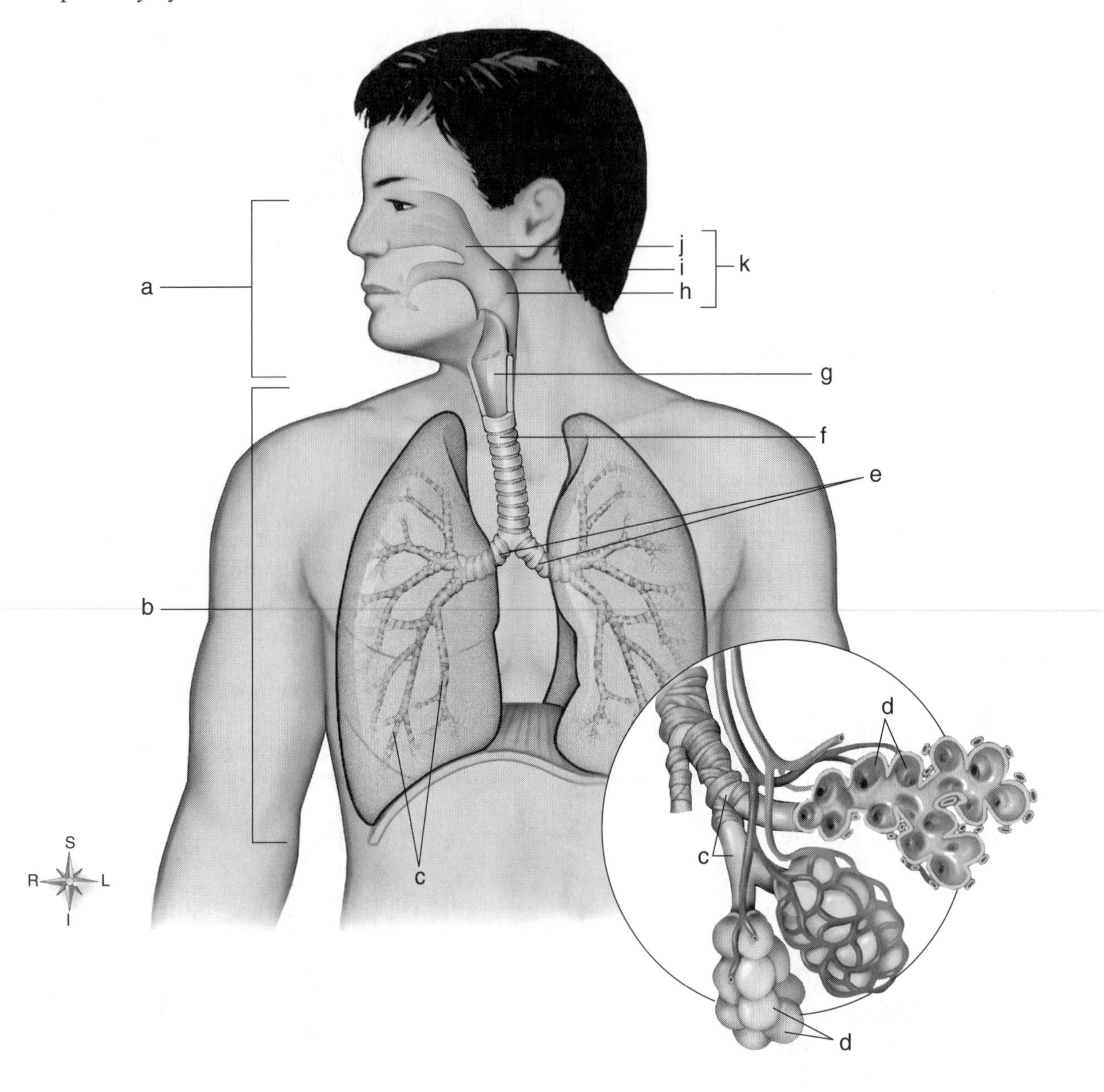

a. ______________________

b. ______________________

c. ______________________

d. ______________________

e. ______________________

f. ______________________

g. ______________________

h. ______________________

i. ______________________

j. ______________________

k. ______________________

59. Lungs and pleura, transverse section

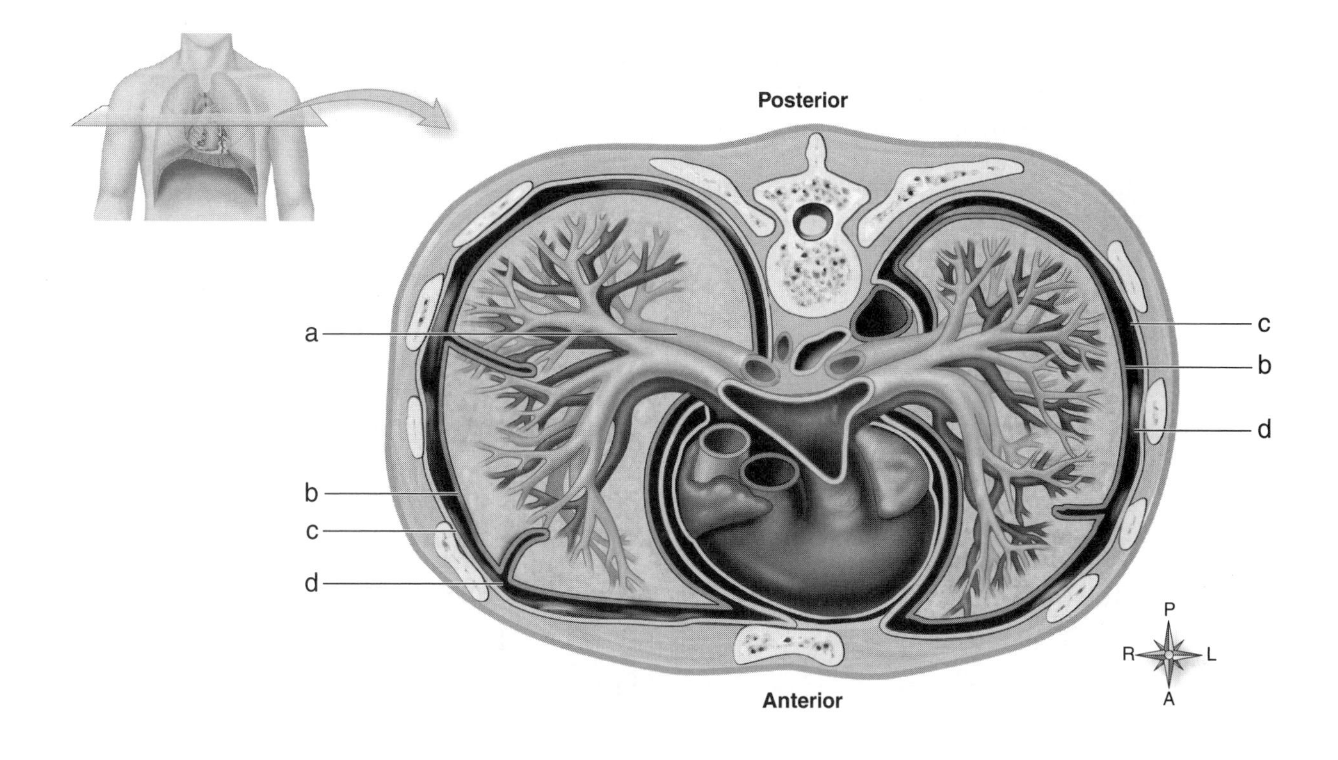

a. ______________________ c. ______________________

b. ______________________ d. ______________________

Short answer

60. List the six structures between the trachea and alveoli.

RESPIRATORY PHYSIOLOGY

STUDY TIP

When it comes to breathing, air ALWAYS moves from positive to negative. If you increase the size of an enclosed cavity (thoracic cavity) you create negative pressure, thus air moves into the lungs. Think of the lungs as balloons. If you put a balloon in a vacuum it gets bigger. To let the air out of the balloon all you have to do is let go and the elastic recoil (elasticity) of the balloon squeezes the air out. If that is not fast enough, you can squeeze the balloon to force the air out quicker (contracting the abdominal muscles and pushing the internal organs up into the diaphragm).

Fill in the blanks with the correct answers.

61. As the diaphragm contracts, the chest cavity ____________________. This creates a pressure ____________________ and the air outside the body is at a(n) ____________________ pressure than the air in the lungs. This in turn causes air to rush ____________________ the lungs. This concept of the flow of air is called the primary ____________________.

62. The exchange of gases in the lungs takes place between ____________________ and blood flowing through ____________________.

63. About 10% of carbon dioxide in the blood travels as ____________________, the rest travels as carbaminohemoglobin and as ____________________.

Matching – match each description, word, or phrase with the correct term.

	Term		Definition
64. ______	Inspiration	a.	"Normal" breathing
65. ______	Total lung capacity	b.	Air you can inhale above normal volume
66. ______	Tidal volume (TV)	c.	IRV + TV + ERV + RV
67. ______	Inspiratory reserve volume (IRV)	d.	Pressure exerted by any one gas
		e.	Diaphragm contracting
68. ______	Vital capacity	f.	Air you can force out
69. ______	Functional residual capacity	g.	ERV + RV
70. ______	Partial pressure	h.	Diaphragm relaxing
71. ______	Residual volume (RV)	i.	Can't exhale
72. ______	Expiratory reserve volume (ERV)	j.	IRV + TV + ERV
73. ______	Expiration		

STUDY TIP

Partial pressure refers to pressure exerted by gases and is used to explain what is happening with regard to diffusion of CO_2 and O_2 in the lungs and blood. Many students get confused with this, but if it gives you trouble, think of it as % rather than P. Example; P_{CO_2} or P_{O_2} vs. $\%CO_2$ or $\%O_2$. Diffusion always goes from high to low. Look at some samples in the book and this might help to clarify.

Multiple choice – select the best answer.

74. Air in the alveoli at the end of one expiration and before the beginning of another inspiration is:
    a. the same as the atmosphere.
    b. greater than the atmosphere.
    c. less than the atmosphere.
    d. pushing the diaphragm down.

FC 75. The primary means of inhalation is:
    a. relaxing the diaphragm.
    b. increasing thoracic pressure.
    c. relaxing external intercostals.
    d. contracting the diaphragm.

FC 76. Exhaling is primarily caused by:
a. relaxing the diaphragm.
b. relaxing the intercostal muscles.
c. elastic recoil.
d. negative interpleuritic pressures.

77. The air moved with normal breathing is called:
a. tidal volume.
b. vital capacity.
c. inspiratory reserve volume.
d. total lung capacity.

78. Which of the following is true?
a. Arterial PO_2 is higher than alveolar PO_2 in the lungs.
b. Arterial PCO_2 is higher than alveolar PCO_2 in the lungs.
c. Alveolar PO_2 is lower than arterial PO_2 in the lungs.
d. Alveolar PCO_2 is higher than arterial PCO_2 in the lungs.

79. Which of the following is correct?
a. CO_2 is moving from the alveoli to the arteries and O_2 is moving from the arteries to the alveoli in the lungs.
b. O_2 and CO_2 are moving from the alveoli to the arteries in the lungs.
c. O_2 and CO_2 are moving from the arteries to the alveoli in the lungs.
d. O_2 is moving from the alveoli to the arteries and CO_2 is moving from the arteries to the alveoli in the lungs.

FC 80. The pressure exerted by a gas in a mixture of gases or liquid is called:
a. total pressure.
b. total partial pressure.
c. percentage pressure.
d. partial pressure.

81. How many layers of cells lie between blood in the vessels in the lungs and the air in the alveoli?
a. 1
b. 2
c. 4
d. 6

FC 82. The main control centers for breathing are located in the:
a. diaphragm.
b. central chemoreceptors in the brainstem.
c. dorsal respiratory control center in the cerebrum.
d. peripheral chemoreceptors in the carotid bodies and aorta.

FC 83. The main "drive" to breathe is:
a. increase of CO_2 or decrease of pH.
b. increase of CO_2 and O_2.
c. increase of pH and decrease of O_2.
d. decrease of O_2.

STUDY TIP

When CO_2 enters the blood it forms carbonic acid. As far as breathing is concerned, CO_2 equals acid (more hydrogen ions). Thus if I breathe faster, I blow off more CO_2 (acid) and my pH goes up. If my breathing slows down (drug overdoses) CO_2 (acid) builds up and my pH goes down. Thus, CO_2 = acid! You may need to review pH in Chapter 2.

Labeling – label the following diagram.

84. Pulmonary ventilation volumes and capacities

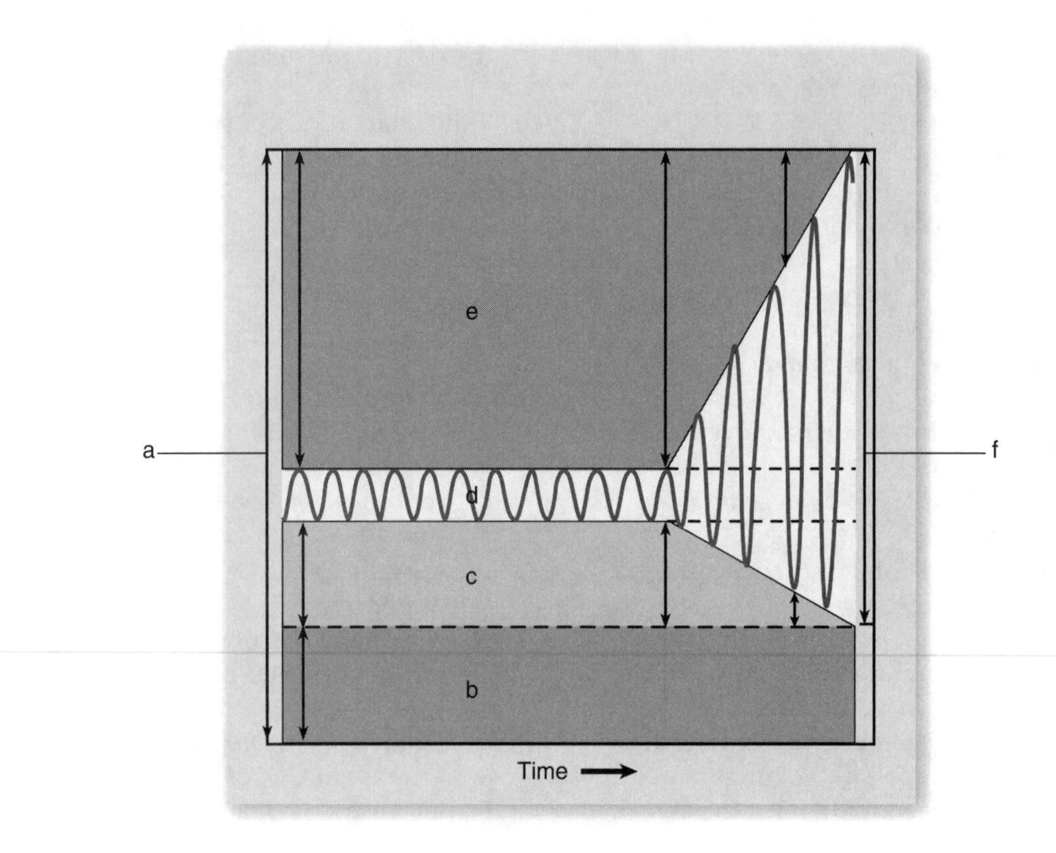

a. ______________________

b. ______________________

c. ______________________

d. ______________________

e. ______________________

f. ______________________

CHAPTER 21

Digestive System

HOW TO APPROACH THE DIGESTIVE SYSTEM

If I had to pick one word to describe the function of the digestive system, it would be *energy*. The entire digestive tract is designed to store, process, and eliminate sources of energy (food). It is similar to the gas tank on a car: a small tube leading to a large tank that holds fuel until it is "digested" and then eliminated out another tube. The digestive tract is simply one long tube, mouth to anus. To test yourself on the structure of the GI system, name every structure in order from mouth to anus. Identify EVERY structure that you pass over, by, or through on the way. Example; lips, teeth, cheek, oral cavity, tongue, uvula, oropharynx, laryngopharynx, esophagus, lower esophageal sphincter (LES), stomach (fundus, body, pyloris, rugae, greater curvature, lesser curvature), pyloric sphincter, duodenum, etc. You simply do this over and over until you have it. Sound simple? It really is that simple; you just have to keep doing it. Use a scratch piece of paper as described previously in Chapter 17.

The function of the digestive system is based around "juices" or digestive enzymes. You know what the tube does and can identify all the structures, mouth to anus. Now you simply need to memorize which "juices" are produced where and what each of them do. You could easily make a simple picture of the GI tract and label where the juices are produced with a very simple explanation.

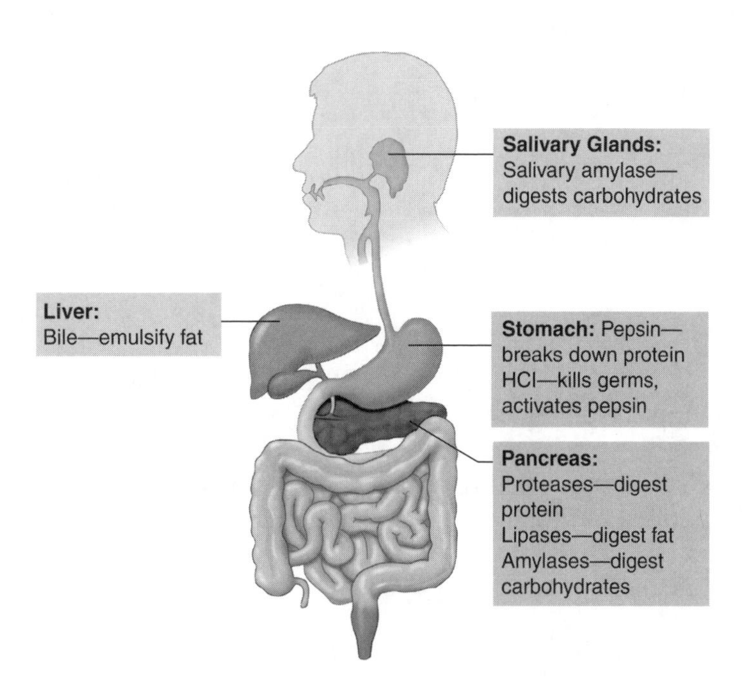

You can do the same thing with the hormones produced throughout the digestive tract. Give it a try. Artistic talent is not required; all you need is to draw a rough picture so you know which organ is which.

MOUTH TO STOMACH

Fill in the blanks with the correct answers.

1. The digestive tract is often referred to as the ____________________ canal or ____________________.

2. The process of chewing food is called ____________________ and swallowing is called ____________________.

3. A rounded mass of food is called a(n) ____________________.

FC *Matching – match each description, word, or phrase with the correct term.*

		Term		Definition
4.	______	Ghrelin	a.	Visceral peritoneum
5.	______	Parietal cells	b.	Largest salivary glands
6.	______	Chief cells	c.	Covers crown
7.	______	Esophageal hiatus	d.	Maxillae and palatine bones
8.	______	Sublingual gland	e.	Produce enzymes
9.	______	Lingual frenulum	f.	Produce hydrochloric acid
10.	______	Hard palate	g.	Frequent location of hernias
11.	______	Serosa	h.	Hormone
12.	______	Parotid	i.	Smallest salivary gland
13.	______	Enamel	j.	Anchors tongue to floor of mouth

Multiple choice – select the best answer.

14. Which of the following is NOT a layer of the GI tract?
    a. Submuscularis
    b. Muscularis
    c. Submucosa
    d. Mucosa

15. Which of the following is true regarding the serosa?
    a. Is actually the visceral peritoneum
    b. Is actually the parietal peritoneum
    c. Produces mucus
    d. Is next to the muscularis

16. The parotid salivary gland produces:
    a. enzymes but not mucus.
    b. mucus only.
    c. enzymes only.
    d. enzymes, mucus, and hormones.

17. The most lateral salivary gland is the:
    a. parotid.
    b. submandibular.
    c. submaxillary.
    d. sublingual.

18. The greatest proportion of the tooth is made up of:
    a. enamel.
    b. dentin.
    c. cementum.
    d. pulp cavity.

19. The outer layer of the root of the tooth is covered by:
    a. bone.
    b. dentin.
    c. enamel.
    d. cementum.

FC 20. The opening in the diaphragm that the esophagus goes through is called the esophageal:
    a. sphincter.
    b. hernia.
    c. hiatus.
    d. canal.

21. The most superior portion of the stomach is the:
    a. body.
    b. fundus
    c. cardia.
    d. pylorus.

FC 22. The expansion ridges or folds of the stomach are called:
    a. rugae.
    b. plicae.
    c. villi.
    d. gastric plicae.

23. Which of the following is NOT a function of the stomach?
    a. Secrete intrinsic factor
    b. Destroy pathogens
    c. Produce majority of the digestive enzymes
    d. Produce acid

24. The most inferior portion of the stomach is the:
    a. fundus.
    b. body.
    c. cardia.
    d. pylorus.

25. What is the function of ghrelin?
    a. Increases appetite
    b. Stimulates production of intrinsic factor
    c. Unravels proteins
    d. Suppresses production of digestive enzymes

Labeling – label the following diagrams.

26. Location of digestive organs

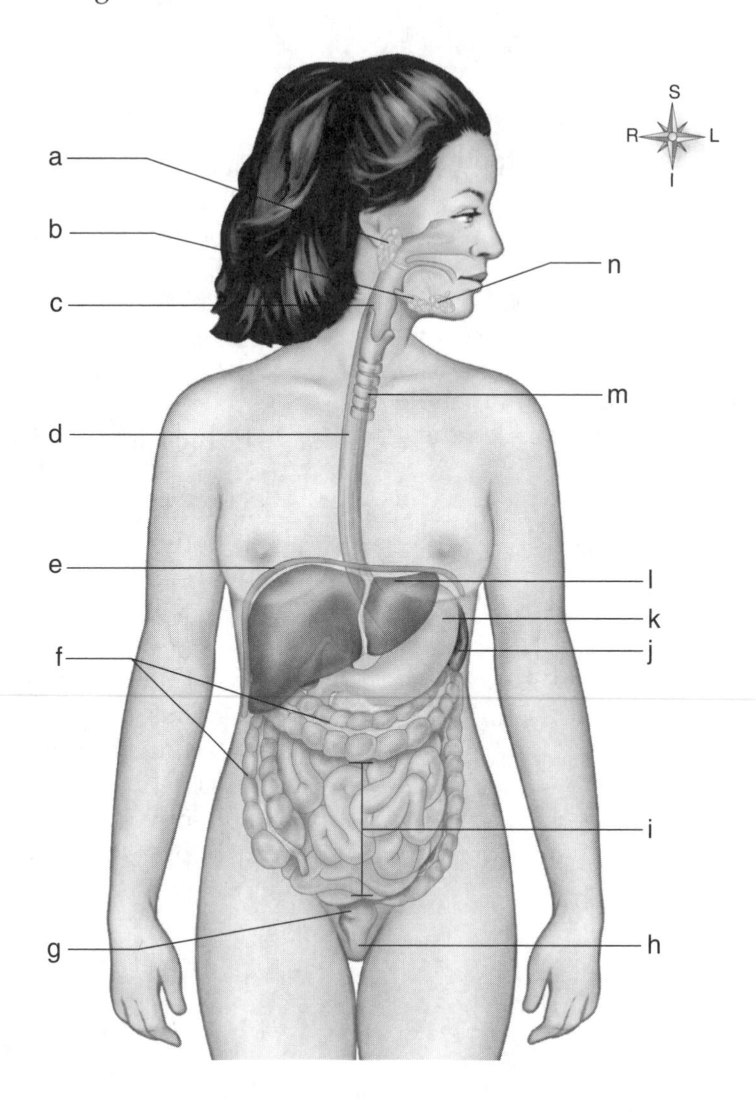

a. ______________________

b. ______________________

c. ______________________

d. ______________________

e. ______________________

f. ______________________

g. ______________________

h. ______________________

i. ______________________

j. ______________________

k. ______________________

l. ______________________

m. ______________________

n. ______________________

27. The oral cavity

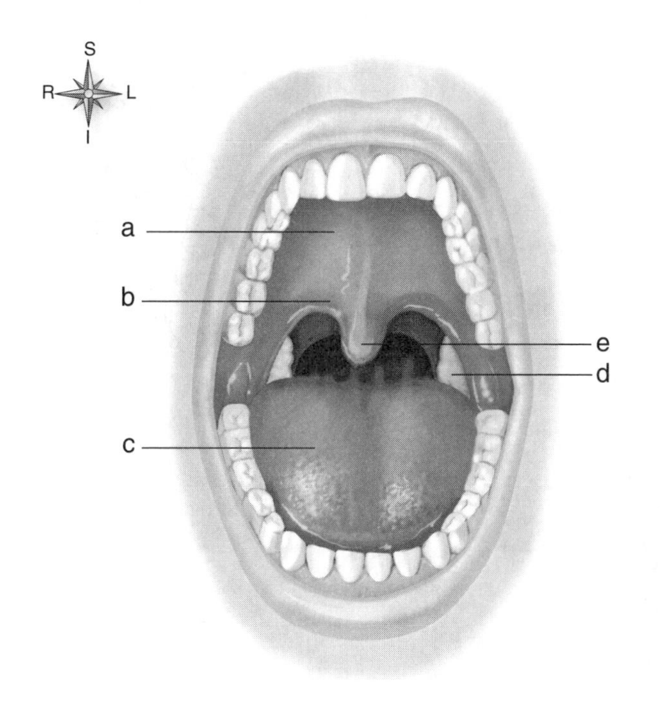

a. ______________________

b. ______________________

c. ______________________

d. ______________________

e. ______________________

28. Typical tooth

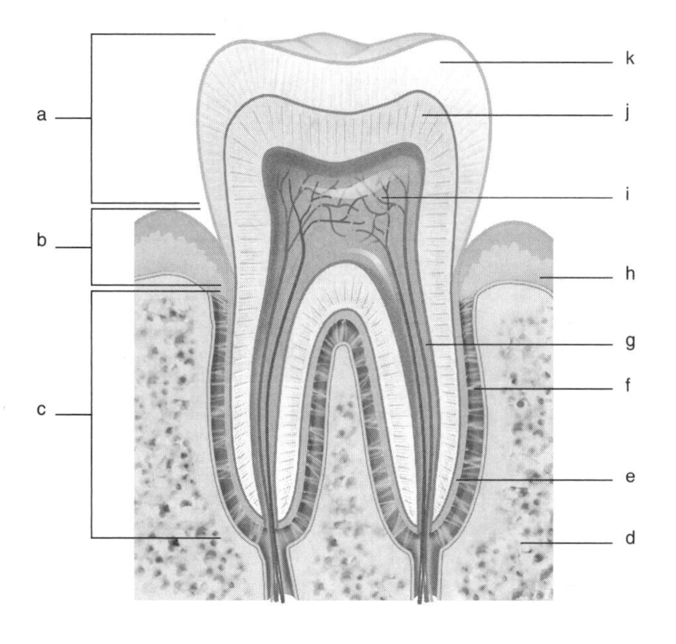

a. ______________________

b. ______________________

c. ______________________

d. ______________________

e. ______________________

f. ______________________

g. ______________________

h. ______________________

i. ______________________

j. ______________________

k. ______________________

29. Stomach

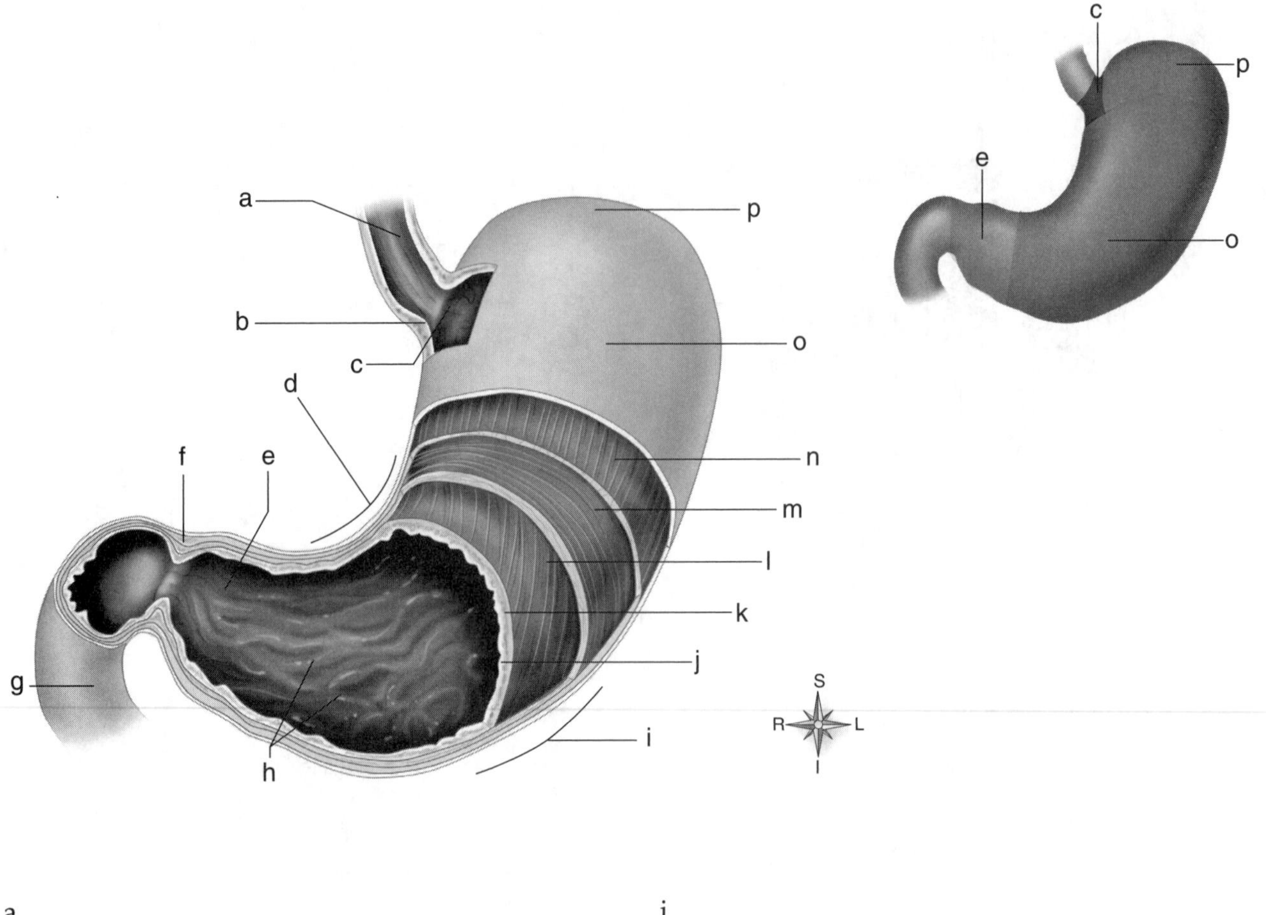

a. ______________________ i. ______________________

b. ______________________ j. ______________________

c. ______________________ k. ______________________

d. ______________________ l. ______________________

e. ______________________ m. ______________________

f. ______________________ n. ______________________

g. ______________________ o. ______________________

h. ______________________ p. ______________________

FC *Short answer*

30. List the four layers of the wall of the GI tract profundus to superficial.

31. Name the four structures that create the oral cavity.

32. Name the three salivary glands medial to lateral.

33. Name the three major parts of a typical tooth.

34. Name the three major divisions of the stomach inferior to superior.

35. Identify four major functions of the stomach.

SMALL INTESTINE TO PANCREAS

FC ***Fill in the blanks with the correct answers.***

36. The majority of digestion takes place in the ____________________.

37. The ____________________ allows free movement of the intestines and helps prevent tangles.

38. The endocrine cells of the pancreas are called ____________________.

FC ***Matching – match each description, word, or phrase with the correct term.***

		Term		Definition
39.	______	Villi	a.	"Fat apron"
40.	______	Taeniae coli	b.	"Pouches"
41.	______	Peritoneum	c.	Contains an arteriole, venule, and lacteal
42.	______	Mesentery	d.	Two layers of parietal peritoneum
43.	______	Plicae	e.	"Folds"
44.	______	Greater omentum	f.	Produce insulin
45.	______	Haustra	g.	Lines the walls of the abdominal cavity
46.	______	Vermiform appendix	h.	Inferior portion of cecum
47.	______	Beta cells	i.	Longitudinal muscles
48.	______	Microvilli	j.	Increases surface area of the cell

Multiple choice – select the best answer.

49. What is the shortest part of the small intestine?
    a. Cecum
    b. Jejunum
    c. Ileum
    d. Duodenum

50. Which of the following is the correct path of feces?
    a. Ileum, cecum, ileocecal valve, ascending colon, splenic flexure
    b. Ascending colon, splenic flexure, transverse colon, hepatic flexure
    c. Cecum, ileocecal valve, ascending colon, splenic flexure, transverse colon
    d. Ileum, ileocecal valve, cecum, ascending colon, hepatic flexure

51. The portion of the large intestine that is positioned vertical and midline is the:
    a. ascending colon.
    b. descending colon.
    c. sigmoid colon.
    d. rectum.

FC 52. Which of the following separates the right and left lobes of the liver?
    a. Falciform ligament
    b. Hepatic ligament
    c. Round ligament
    d. Broad ligament

53. The greater omentum attaches where?
    a. Greater omentum arch
    b. Greater curvature
    c. Gastric curvature
    d. Greater gastric curvature

54. The functional anatomical units of the liver are the hepatic:
    a. lobules.
    b. units.
    c. papilla.
    d. ducts.

FC 55. Which of the following is NOT a main component of bile?
    a. Bile salts
    b. Bile pigments
    c. Cholesterol
    d. Bile acids

56. Which of the following is true regarding the gallbladder?
    a. It concentrates bile.
    b. It produces bile.
    c. Is directly connected to the common hepatic duct.
    d. Stores digestive enzymes along with bile.

FC 57. The organ that produces the majority of digestive enzymes is the:
    a. stomach.
    b. liver.
    c. pancreas.
    d. gallbladder.

Labeling – label the following diagrams.

58. Wall of the small intestine

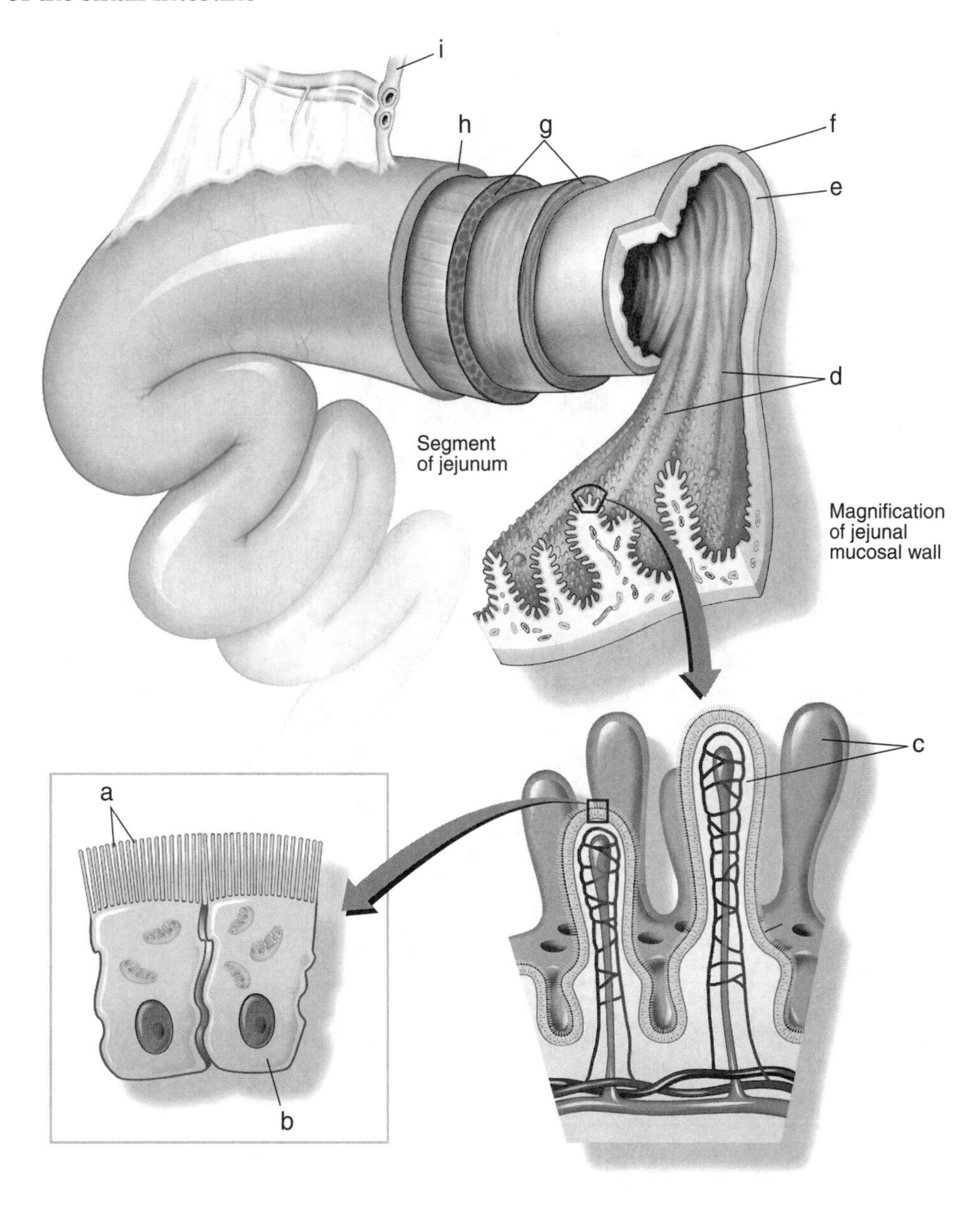

a. ______________________

b. ______________________

c. ______________________

d. ______________________

e. ______________________

f. ______________________

g. ______________________

h. ______________________

i. ______________________

59. Divisions of the large intestine

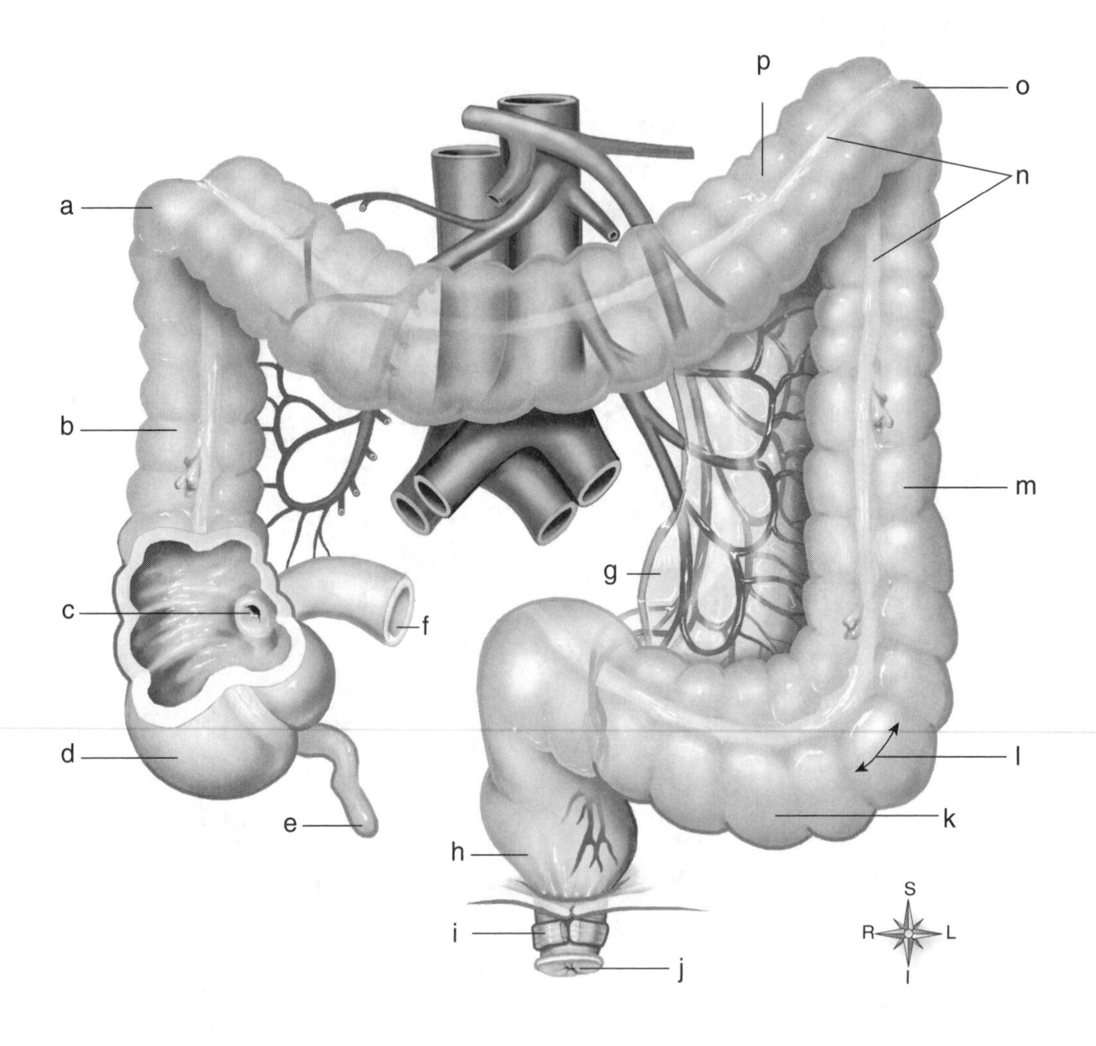

a.	__________	i.	__________
b.	__________	j.	__________
c.	__________	k.	__________
d.	__________	l.	__________
e.	__________	m.	__________
f.	__________	n.	__________
g.	__________	o.	__________
h.	__________	p.	__________

60. The rectum and anus

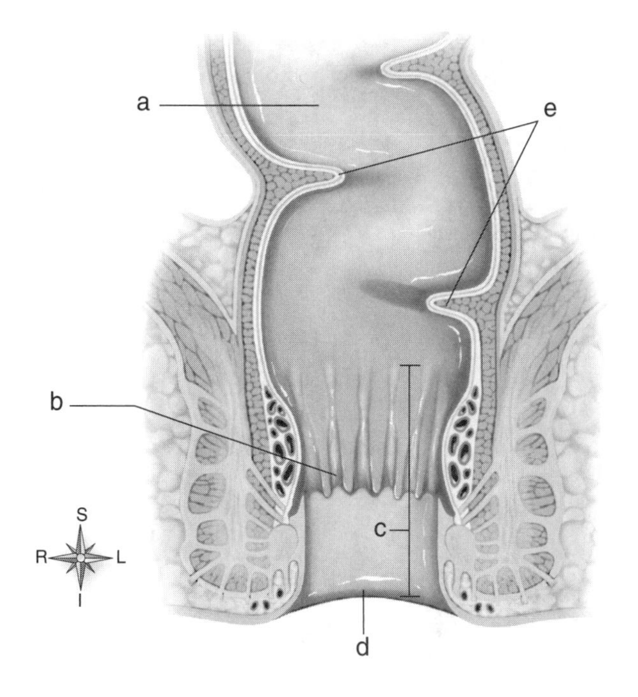

a. ______________________

b. ______________________

c. ______________________

d. ______________________

e. ______________________

61. Structure of the liver

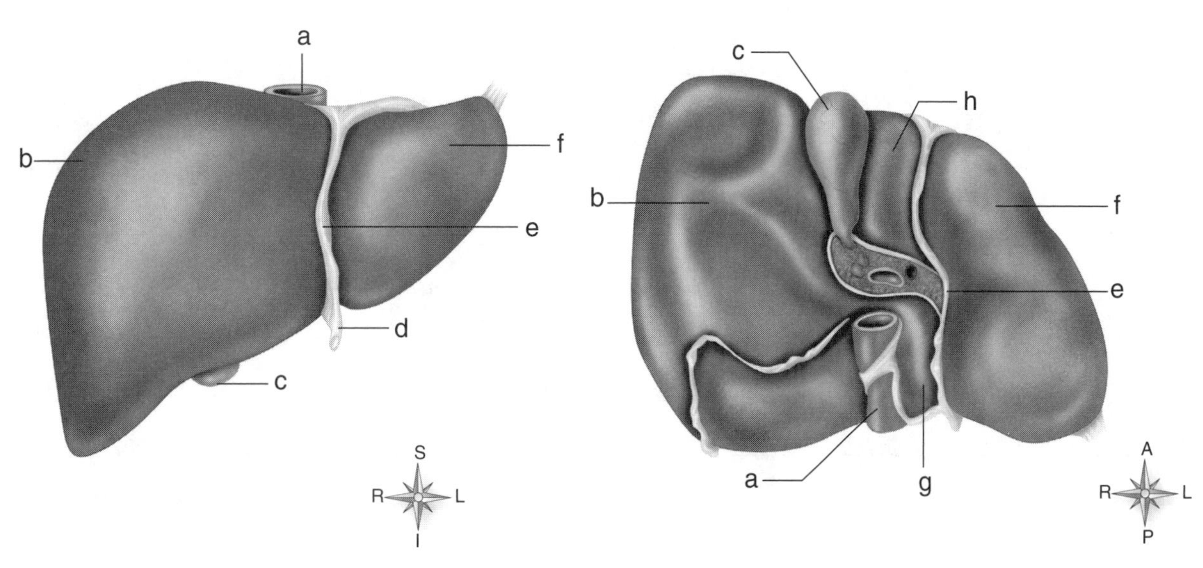

a. ______________________

b. ______________________

c. ______________________

d. ______________________

e. ______________________

f. ______________________

g. ______________________

h. ______________________

62. Bile ducts

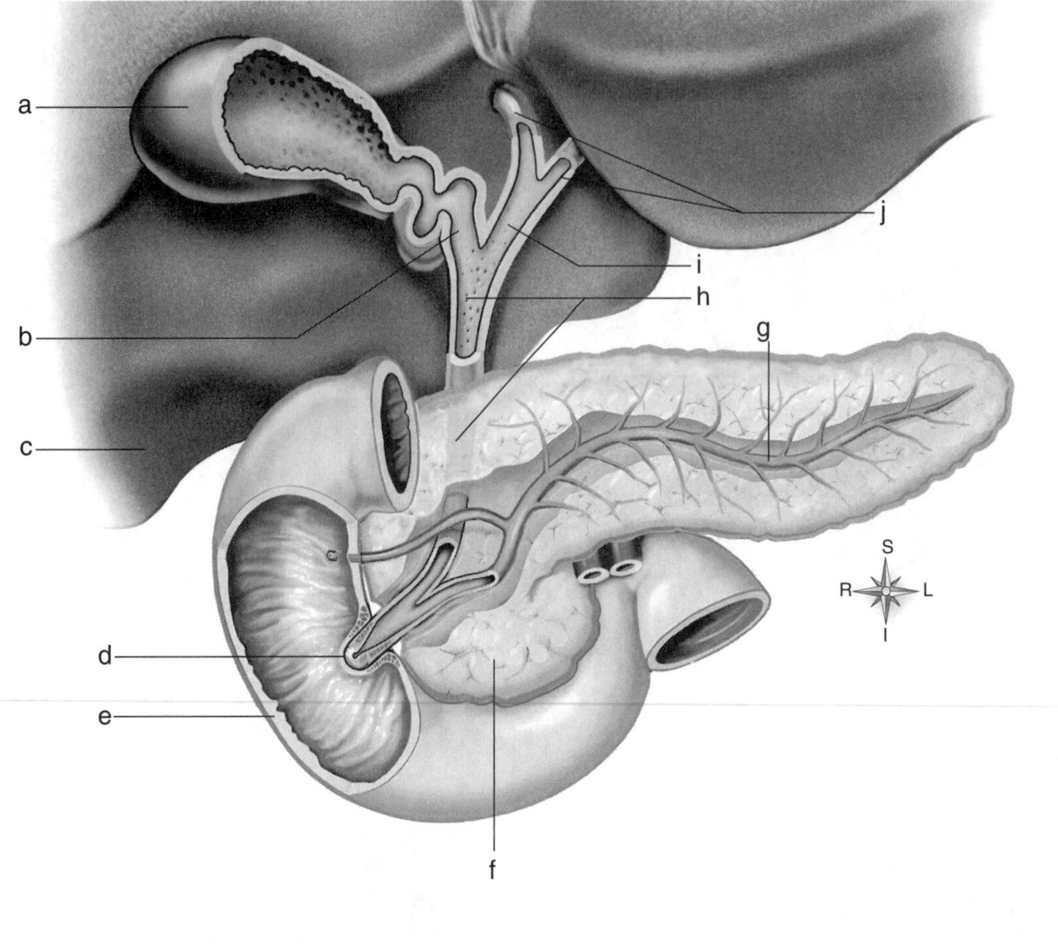

a. ______________________________________
b. ______________________________________
c. ______________________________________
d. ______________________________________
e. ______________________________________
f. ______________________________________
g. ______________________________________
h. ______________________________________
i. ______________________________________
j. ______________________________________

FC *Short answer*

63. Name the three divisions of the small intestine inferior to superior.

64. Identify the nine divisions/structures of the large intestine in order from right after ilium to anus.

65. Name the four lobes of the liver.

66. List four functions of the liver.

67. Name the three types of cells of the pancreas.

DIGESTIVE FUNCTION

Fill in the blanks with the correct answers.

68. ____________________ consists of mastication, deglutition, peristalsis, and segmentation.

FC 69. ____________________ is often described as a wavelike motion of materials through the GI tract.

70. ____________________ is the mixing movement of materials in the GI tract.

71. ____________________ is the chemical transformation of food.

FC 72. ____________________ break down carbohydrates, ____________________ break down proteins, and ____________________ break down lipids.

FC ***Matching – match each description, word, or phrase with the correct term.***

	Term		Definition
73. _____	Cholecystokinin (CCK)	a.	Emulsification
74. _____	Secretin	b.	Protects vitamin B_{12} from digestion
75. _____	Bile	c.	Converts pepsinogen to pepsin
76. _____	Hydrochloric acid (HCl)	d.	Inhibits HCl production
77. _____	Intrinsic factor	e.	Converts proenzymes to active enzymes
78. _____	Pepsin	f.	Digest RNA and DNA
79. _____	Trypsin	g.	Stimulates pancreas to release enzymes
80. _____	Nucleases	h.	Protein-digesting enzyme from pancreas
81. _____	Amylases	i.	Produced in stomach and unravels proteins
82. _____	Kinases	j.	Digests carbohydrates

Multiple choice – select the best answer.

83. The chemical process in which compounds unite with water and then split into simpler compounds is called:
    a. segmentation.
    b. hydrolysis.
    c. mechanical digestion.
    d. emulsification.

FC 84. Proenzymes:
    a. are digestive enzymes synthesized in an inactive form.
    b. activate most digestive enzymes.
    c. combine with micelles to form kinases.
    d. are hormones that stimulate enzyme production.

85. Which of the following is NOT a protease?
    a. Pepsin
    b. Trypsin
    c. Chymotrypsin
    d. CCK

86. Lecithin mixes with lipids and water forming tiny droplets of fat called:
    a. lipases.
    b. bile salts.
    c. micelles.
    d. zymogens.

FC 87. Breaking large drops of fat into smaller drops of fat is called:
    a. emulsification.
    b. digestion.
    c. mechanical digestion.
    d. hydrolysis.

FC 88. Chief cells in the stomach produce what?
    a. Pepsinogen
    b. Pepsin
    c. HCl
    d. Intrinsic factor

FC 89. Parietal cells in the stomach produce what?
    a. HCl and intrinsic factor
    b. HCl and pepsinogen
    c. HCl
    d. Intrinsic factor and HCl

FC 90. Which of the following hormones stimulates release of HCO_3 from the pancreas?
    a. CCK
    b. Secretin
    c. Ghrelin
    d. Gastrin

FC 91. Which of the following stimulates the gallbladder to contract?
    a. CCK
    b. HCl
    c. Secretin
    d. HCO_3

92. Acinar cells are located in the:
    a. stomach.
    b. liver.
    c. pancreas.
    d. intestine.

FC 93. Acinar cells produce what?
    a. Bile
    b. Glucagon
    c. Intrinsic factor
    d. Digestive enzymes

94. Inactive proenzymes are called:
    a. zymogens.
    b. prozymes.
    c. enzyamases.
    d. proteases.

FC 95. Which of the following hormones are produced by the intestinal mucosa?
    a. Secretin
    b. CCK
    c. Ghrelin, secretin, CCK
    d. CCK and secretin

STUDY TIP

There are several hormones and enzymes produced along the digestive tract. They can become very confusing. Another way to keep them straight would be to list them under each of the organs along with a VERY simple definition. Example:

LIVER

Bile – emulsify fat

STOMACH

HCl – acid, kill pathogens, activate pepsinogen to pepsin (made by parietal cells)
Pepsinogen – converts to pepsin – protein digestive enzyme (made by chief cells)
Etc…

You may even want to put them in a simple chart.

Short answer

FC 96. List the three amylases.

FC 97. List the four main proteases.

98. Name the three phases of gastric secretions.

99. Identify the three stages of deglutition.

FC 100. Name the three types of enzymes produced by the pancreas.

CHAPTER 22

Nutrition and Metabolism

HOW TO APPROACH NUTRITION AND METABOLISM

Before beginning this chapter, you should review Chapter 2. The main focus of this chapter is energy processing and production. This starts with glycolysis and ends with the electron transport system. The end result of these processes is ATP, the fuel our bodies need to work. Different phases produce different amounts of ATP and different waste products. Students frequently find the process of metabolism difficult, which is why I strongly recommend making your own diagram of the processes. Use the text to help you put it into your words and in a manner that makes sense to you. It will not take a lot of work nor artistic talent. Here is an example. This should help get you started and you should wait until you have heard the lecture covering the material so you know how simple or complex to make your diagram.

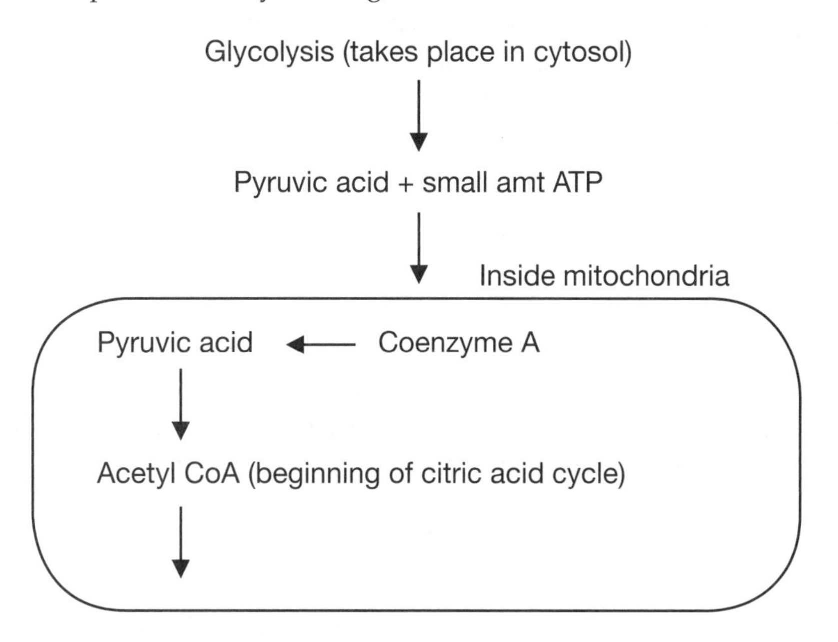

Fill in the blanks with the correct answers.

FC 1. _____________________ refers to the foods we eat and the types of nutrients they contain and _____________________ refers to how we use the food.

2. Glucose reacting with ATP to form a new molecule that cannot move back across the cell membrane is called ___________________. The process of adding a phosphate group to a molecule is called ___________________.

FC 3. The ___________________ is a series of chemical reactions that takes place in the mitochondria and produces large quantities of high-energy electrons.

4. ___________________ are a class of organic compounds that include fats and oils.

FC 5. Body weight will ___________________ when energy input exceeds energy output and ___________________ when energy input is less than energy output.

FC ***Matching – match each description, word, or phrase with the correct term.***

	Term		**Definition**
6. ______	Macronutrients	a.	"Building"
7. ______	Micronutrients	b.	Creates pyruvic acid molecules in the cytoplasm
8. ______	Catabolism	c.	Need in large amounts
9. ______	Anabolism	d.	Building blocks of proteins
10. ______	Glycolysis	e.	Smallest amount of energy expenditure to maintain life, consciousness, and normal body temperature
11. ______	Citric acid cycle	f.	Need in small amounts
12. ______	Lipoproteins	g.	Fat covered with protein shell
13. ______	Amino acids	h.	Takes place in the mitochondria and produces large quantities of ATP
14. ______	Basal metabolic rate	i.	Inorganic elements or salts
15. ______	Minerals	j.	"Destroying"

STUDY TIP

Think of lipoproteins as M&Ms. A fatty center with a water-soluble coating that "melts in your mouth." Cholesterol is not soluble in water (your blood), so it is coated with protein that is soluble in water, thus it can now move through the blood.

Multiple choice – select the best answer.

16. Which of the following will NOT be digested?
    a. Polysaccharides
    b. Glycogen
    c. Cellulose
    d. Disaccharides

FC 17. Glucose is a:
    a. monosaccharide.
    b. disaccharide.
    c. polysaccharide.
    d. starch.

FC 18. The first step in carbohydrate catabolism is called:
   a. citric acid cycle.
   b. Krebs cycle.
   c. glycolysis.
   d. phosphorylation.

FC 19. Which of the following only takes place in the cytoplasm and produces two pyruvic acid molecules and a small amount of ATP?
   a. Glycolysis
   b. Citric acid cycle
   c. Oxidative phosphorylation
   d. ATP synthase

20. Pyruvic acid molecules combine with coenzyme A to form:
   a. glucose.
   b. citric acid.
   c. acetyl CoA.
   d. high-energy electrons.

21. Pyruvic acid will generate which of the following while going through the citric acid cycle?
   a. CO_2, some ATP, and many high-energy electrons
   b. CO_2, H_2O, large amounts of ATP
   c. CO_2 and H_2O
   d. ATP and H_2O

FC 22. Which cycle or process produces the largest amount of ATP?
   a. Glycolysis
   b. Electron transport system
   c. Citric acid cycle
   d. Oxidative phosphorylation

23. Nicotinamide adenine dinucleotide (NAD) and flavin adenine dinucleotide (FAD) are what?
   a. Vitamins
   b. Minerals
   c. Carrier molecules
   d. Enzymes

24. What binds together ADP and a phosphate group to generate ATP?
   a. ATP synthase
   b. ATP coenzyme
   c. NAD and FAD
   d. ATPase

25. What role does oxygen play in oxidative phosphorylation?
   a. Creates more ATP by using up excess high-energy electrons
   b. Serves as an "electron dump," oxidizing the hydrogen into water
   c. Uses proton movement to bind together ADP
   d. Combines with oxaloacetic acid to form citric acid

26. Which of the following is NOT a common lipid in our diet?
   a. Triglycerides
   b. Glycerol
   c. Glycogen
   d. Fatty acids

27. Which of the following is true regarding saturated fats?
   a. They are typically solid at room temperature.
   b. They contain fatty acid chains with double bonds.
   c. They cannot broken down any further.
   d. They are only found in animal products.

28. Small fat droplets found in the blood soon after fat absorption takes place are called:
   a. chylomicrons.
   b. free fatty acids.
   c. lipoproteins.
   d. unsaturated fats.

FC 29. The "good" or "healthy" cholesterol is:
   a. LDL.
   b. VLDL.
   c. HDL.
   d. HLDL.

FC 30. Which of the following is our main form of fat storage?
a. Cholesterol
b. Phospholipids
c. Triglycerides
d. Free fatty acids

FC 31. How many different amino acids do we need?
a. 5
b. 10
c. 15
d. 20

32. Protein metabolism is called ________ and produces ________ as a waste product.
a. ketosis, CO_2
b. deamination, ammonia
c. proteases, ketones
d. protein oxidation, amino acids

33. Which of the following is NOT true regarding vitamins?
a. Most attach to enzymes or coenzymes.
b. They help enzymes work properly.
c. They are fat-soluble or water-soluble.
d. They are inorganic elements.

34. Which of the following is NOT true regarding minerals?
a. Are organic molecules
b. Necessary for nerve impulse conduction
c. Attach to some enzymes to help them work
d. Used in body fluid management

FC *Short answer*

35. What are the two nutrient categories?

36. Name the three forms of carbohydrates.

37. List three methods of transporting lipids in the blood.

38. List the three types of lipoproteins.

39. Name the four most common types of lipids in our body.

40. Name four hormones that control lipid metabolism.

41. Name six things that affect metabolic rate.

42. Name five macronutrients.

43. Name two micronutrients.

CHAPTER 23

Urinary System and Fluid Balance

HOW TO APPROACH THE URINARY SYSTEM AND FLUID BALANCE

The purpose of the urinary system ranges from control of blood production to fluid and electrolyte balance. Fluid balance or urine production is our main focus. Urine comes from blood that has been filtered; thus to begin understanding HOW we make urine, we must first learn all the structures or "plumbing." First learn the basic structures of the kidney—this is the "plumbing" system that is going collect urine. In order to make urine, we need lots of blood. That is the second task, to make sure you can trace blood from renal artery (in) to vein (out) and everything in between. Again, use a scratch piece of paper as previously discussed in Chapter 17.

Now that you have learned the basic structures and the blood supply, you can start focusing on the nephron. The nephron is the unit that connects the blood to the "plumbing;" in other words, this is where urine is made. Blood goes into the nephron and urine comes out. This gives you three major sections: blood supply, nephron, plumbing. This is by far the greatest challenge of this chapter. One you have this concept down, other functions of the kidney and fluid balance fall into place. I can't emphasize enough to trace the "flow" of things on paper over and over! See the study tips below.

ANATOMY OF THE URINARY SYSTEM

Fill in the blanks with the correct answers.

FC 1. The kidneys lie ____________________ against the posterior wall of the abdomen.

2. The ureters and the bladder are lined with ____________________, which permits significant stretching.

3. The ureter openings lie at the posterior corners of the triangle-shaped floor called the ____________________.

4. The many pores present in the glomerular endothelium are called ____________________.

FC 5. The majority of nephrons are ____________________ nephrons, the remainder are called ____________________ nephrons.

FC ***Matching – match each description, word, or phrase with the correct structure.***

	Structure		Definition
6. ______	Hilum	a.	Principal organ of urinary system
7. ______	Renal pyramid	b.	Where everything enters or leaves the kidney
8. ______	Renal cortex	c.	Outer layer
9. ______	Renal columns	d.	Inner layer
10. ______	Minor calyces	e.	Located in the medulla
11. ______	Kidney	f.	Cortex located in the medulla
12. ______	Renal medulla	g.	Drains the renal papillae
13. ______	Major calyces	h.	Drain minor calyces
14. ______	Renal pelvis	i.	Transports urine to bladder
15. ______	Ureters	j.	Tip of pyramid
16. ______	Renal papilla	k.	Large funnel that collects urine to leave kidney

FC ***Matching – match each description, word, or phrase with the correct term.***

	Term		Definition
17. ______	Renal corpuscle	a.	Consists of a thick and thin limb
18. ______	Bowman capsule	b.	Connects to Bowman capsule
19. ______	Filtrate	c.	Secretes renin and helps regulates BP
20. ______	Filtration slits	d.	Bowman capsule and glomerulus
21. ______	Glomerulus	e.	Formed by podocytes
22. ______	Proximal convoluted tubule (PCT)	f.	What "leaks" out of glomerulus
23. ______	Nephron	g.	Collects filtrate
24. ______	Henle loop	h.	Epithelial cells with "feet-like" projections
25. ______	Juxtaglomerular (JG) apparatus	i.	Joins several nephrons and opens into minor calyx
26. ______	Podocyte	j.	"Leaky" capillary network
27. ______	Collecting duct (CD)	k.	PCT, Henle loop, distal convoluted tubule (DCT), and CD

STUDY TIP

Remember that when it comes to sphincters for both renal and digestive, Internal = Involuntary. That means that external is voluntary. Just remember I = I.

Multiple choice – select the best answer.

28. Which of the following protrudes into the minor calyces?
    a. Renal pelvis
    b. Renal column
    c. Renal papilla
    d. Major calyces

29. Which of the following is the correct flow of blood in the kidney?
    a. Renal artery, segmental artery, lobar artery, interlobar artery, arcuate artery, interlobular artery
    b. Renal artery, lobar artery, segmental artery, interlobar artery, arcuate artery, interlobular artery
    c. Renal artery, segmental artery, lobar artery, interlobar artery, interlobular artery, arcuate artery
    d. Renal artery, segmental artery, lobar artery, interlobular artery, arcuate artery, interlobar artery

FC 30. Which blood vessel goes into the renal corpuscle?
    a. Efferent
    b. Arcuate
    c. Renal
    d. Afferent

31. Which of the following is the correct flow of urine/filtrate?
    a. Glomerular capsule, PCT, Henle loop, DCT, minor calyces, major calyces
    b. Renal papillae, minor calyces, major calyces, renal pelvis, ureter, bladder
    c. DCT, minor calyces, major calyces, renal sinus, renal pelvis, ureter
    d. DCT, Henle loop, PCT, CD, minor calyces, major calyces, renal pelvis

FC 32. The functional unit of the kidney is called the:
    a. renal corpuscle.
    b. nephron.
    c. renal pyramid.
    d. renal tubule.

33. The muscles of the bladder are collectively called the:
    a. detrusor muscle.
    b. renal muscle.
    c. cystic muscle.
    d. trigone muscle.

34. Which of the following muscles is completely involuntary?
    a. Detrusor
    b. Internal urethral sphincter
    c. External urethral sphincters
    d. Levator ani

35. Which of the following is NOT part of the nephron?
    a. Renal corpuscle
    b. Henle loop
    c. Collecting duct
    d. Bowman capsule

36. The visceral wall of Bowman capsule is composed of special epithelial cells called:
    a. Bowman cells.
    b. podocytes.
    c. epithelial filtrate cells.
    d. pedicelcytes.

37. The JG apparatus secretes what in response to low blood pressure?
    a. Angiotensin I
    b. Angiotensin II
    c. Renin
    d. Aldosterone

38. Ions are reabsorbed from the renal tubule back into the vasa recta or the:
    a. arcuate vein.
    b. afferent artery.
    c. efferent artery.
    d. peritubular capillaries.

Labeling – label the following diagrams.

39. Location of the urinary system organs

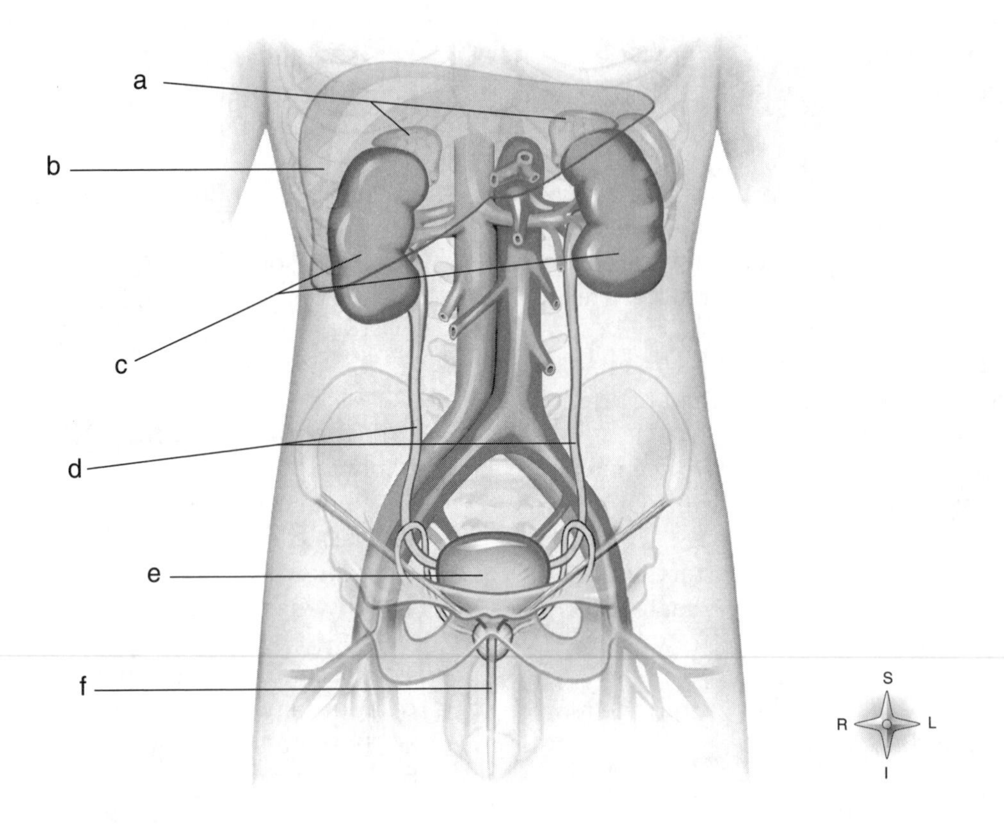

a. ______________________

b. ______________________

c. ______________________

d. ______________________

e. ______________________

f. ______________________

40. Internal structure of the kidney

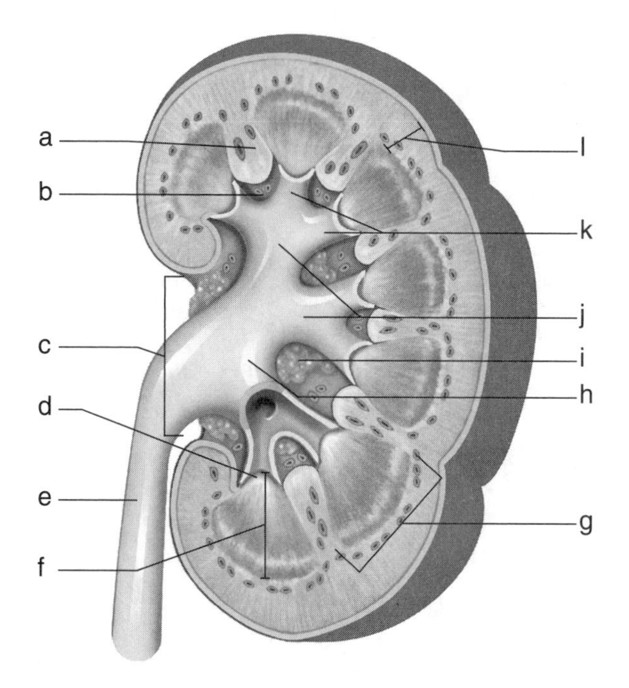

a. ______________________

b. ______________________

c. ______________________

d. ______________________

e. ______________________

f. ______________________

g. ______________________

h. ______________________

i. ______________________

j. ______________________

k. ______________________

l. ______________________

41. Circulation of blood through the kidney

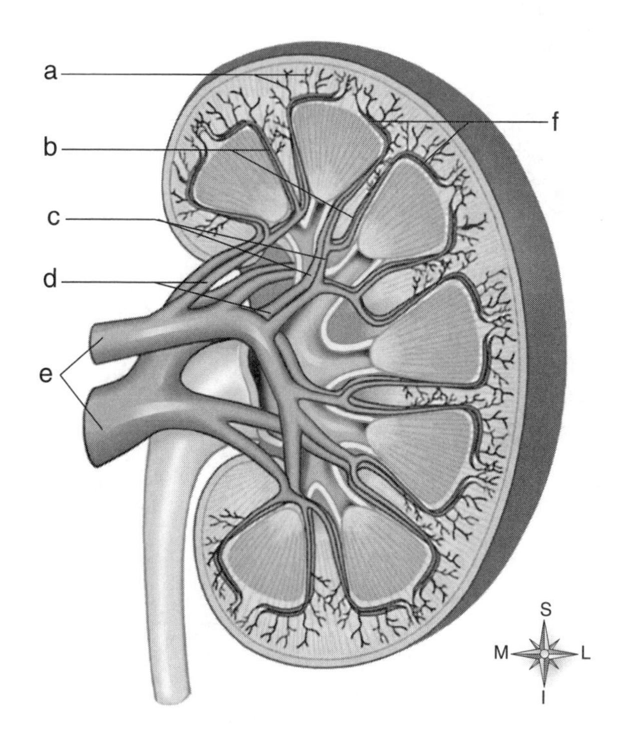

a. ______________________

b. ______________________

c. ______________________

d. ______________________

e. ______________________

f. ______________________

42. Nephron

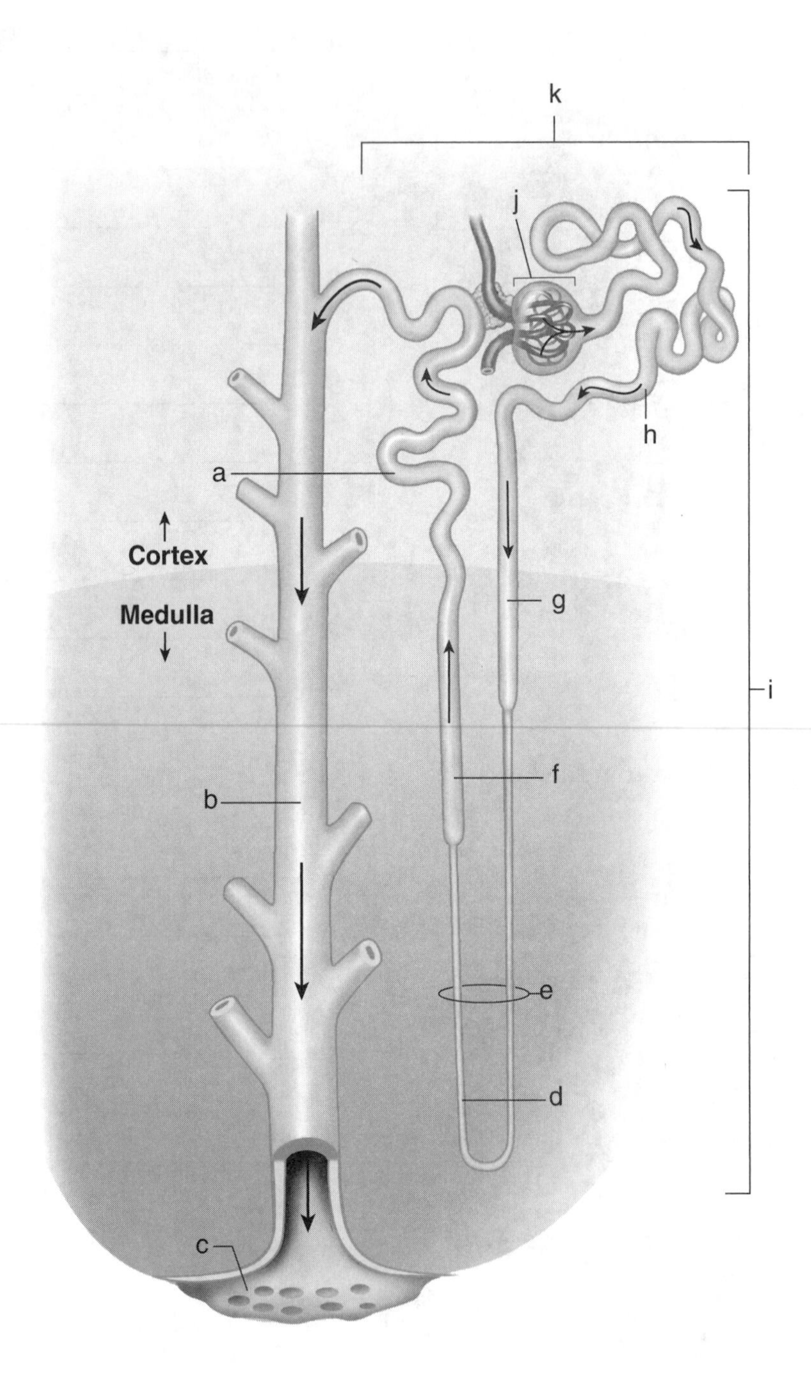

a. ______________________

b. ______________________

c. ______________________

d. ______________________

e. ______________________

f. ______________________

g. ______________________

h. ______________________

i. ______________________

j. ______________________

k. ______________________

43. Structure of renal corpuscle

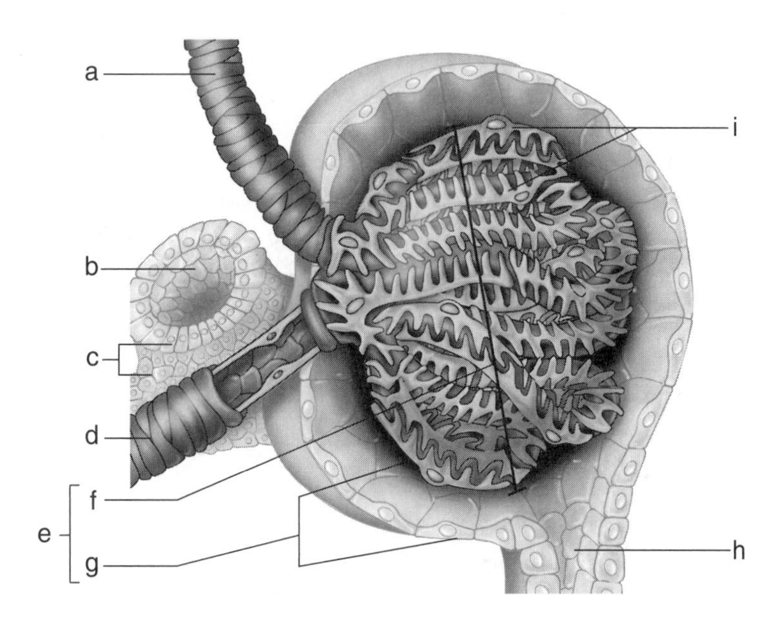

a. ______________________

b. ______________________

c. ______________________

d. ______________________

e. ______________________

f. ______________________

g. ______________________

h. ______________________

i. ______________________

FC *Short answer*

44. Name the two components of the renal corpuscle.

45. Name the three sections of the renal tubule.

46. Name the two different types of nephrons.

47. Name the four parts of the nephron.

STUDY TIP

At this point you should be able to trace blood all the way through the kidney. Another helpful test would be to trace a molecule of urea that is going to get filtered out in the nephron and end up in the toilet. Using your scratch paper that you have handy and start at the renal artery and make your way to the glomerulus where our urea molecule is filtered out in the capsule and makes it way to PCT. Keep going all the way to the toilet. If I have not said this before, when doing this on paper all you have to do is abbreviate as short as possible as long as you know what you meant. Speed is the name of this game. Example:

RA (renal artery) – seg. A – LA (lobar artery) – ILA – A – Ilobular – Aff – G – BC (bowman capsule) – PCT – get it? You can also do this out loud with a study partner or group. You can even have family and friends help if you have written it out for them to read as you recite the order. If you can do this with 100% accuracy, you will do well on exams.

PHYSIOLOGY OF THE URINARY SYSTEM

FC *Matching – match each description, word, or phrase with the correct term.*

	Term		Definition
48. ______	Sodium cotransport	a.	"Opposite directions"
49. ______	Tubular secretion	b.	Promotes secretion of sodium into urine/filtrate
50. ______	Countercurrent mechanism	c.	Promotes reabsorption of sodium from the urine/filtrate
51. ______	Tubular reabsorption	d.	Movement of water and solutes from blood to capsular space
52. ______	Antidiuretic hormone (ADH)	e.	Causes collecting ducts to become more permeable to water
53. ______	Filtration	f.	Takes place in the PCT and primarily involves reabsorption of glucose and amino acids
54. ______	Aldosterone	g.	Movement of molecules from nephron back into blood
55. ______	Atrial natriuretic hormone (ANH)	h.	Movement of molecules from blood into renal tubule
56. ______	Renin	i.	Leads to production of aldosterone

Multiple choice – select the best answer.

57. Movement of water and protein-free solutes from the glomerulus to the capsular space is called:
    a. tubular reabsorption.
    b. glomerular secretion.
    c. tubular secretion.
    d. filtration.

FC 58. What is needed to push fluid out of the glomerulus?
    a. Hydrostatic pressure
    b. Osmotic pressure
    c. Filtration pressure
    d. Negative pressure gradient

FC 59. The majority of the filtrate is reabsorbed in the:
a. CD.
b. DCT.
c. Henle loop.
d. PCT.

60. Glucose and amino acids are reabsorbed by means of:
a. sodium cotransport mechanism.
b. reverse filtration.
c. osmosis.
d. diffusion.

FC 61. What is the main form of nitrogen-containing waste found in urine?
a. Ammonia
b. Amino acids
c. Urea
d. Uric acid

FC 62. The majority of the solutes in the filtrate are reabsorbed in the:
a. PCT.
b. DCT.
c. CD.
d. Henle loop.

63. ADH targets what structure(s) to become more permeable to water?
a. PCT and DCT
b. DCT only
c. PCT only
d. DCT and CD

FC 64. Which of the following is true regarding aldosterone?
a. Secreted by the JG apparatus
b. Increases reabsorption of sodium
c. Stimulates release of renin
d. Decreases blood pressure

FC 65. Which of the following is true regarding ANH?
a. Inhibits natriuresis
b. Enhances the effects of aldosterone
c. Promotes excretion of sodium
d. Is secreted in response to production of renin

FC *Short answer*

66. List the three processes by which urine is formed.

67. List four functions of the kidney.

68. Name four hormones that affect urine production.

69. List five substances excreted in urine.

FLUID AND ELECTROLYTE BALANCE

Fill in the blanks with the correct answers.

70. The more hydrogen ions present, the ____________________ the pH; the less hydrogen ions present, the ____________________ the pH.

71. The two major fluid compartments are ____________________ and ____________________.

72. ____________________ make up the functional thirst center of the brain.

FC ***Matching – match each description, word, or phrase with the correct term.***

	Term		Definition
73. ______	Extracellular fluid (ECF)	a.	Measurement of hydrogen ion concentration
74. ______	Intracellular fluid (ICF)	b.	Positive ions
75. ______	Interstitial fluid	c.	Majority of our body water
76. ______	Cations	d.	Includes intravascular and interstitial fluid
77. ______	Anions	e.	"Too much fluid"
78. ______	Osmoreceptors	f.	Negative ions
79. ______	Edema	g.	Detect increase in solute concentration
80. ______	pH	h.	Fluid that surrounds cells
81. ______	Acidosis	i.	Arterial blood pH greater than 7.45
82. ______	Alkalosis	j.	Arterial blood pH less than 7.35

STUDY TIP

REMEMBER – water always follows sodium. If you conserve sodium, you conserve water and vice versa. You are so good at keeping your sodium that if your sodium is high or low it is actually a problem of too much or too little water. A nephrologist once taught me it's NOT a salt problem, IT'S A WATER PROBLEM. Think about it—if you have too much water, what would that do to your sodium concentration? It would dilute it, make it appear low. Get it?

Multiple choice – select the best answer.

83. Which of the following is NOT a cation?
    a. Sodium
    b. Calcium
    c. Chloride
    d. Magnesium

FC 84. What is the number-one intracellular cation?
    a. Sodium
    b. Potassium
    c. Chloride
    d. Calcium

FC 85. What is the number-one extracellular cation?
   a. Sodium
   b. Potassium
   c. Calcium
   d. Chloride

86. What does plasma contain that interstitial fluid does not or has very little of?
   a. Electrolytes
   b. Protein
   c. Urea
   d. Water

87. What is the most important extracellular anion?
   a. Potassium
   b. Chloride
   c. Hydrogen
   d. Hydroxide

FC 88. If sodium is reabsorbed, what else will be reabsorbed?
   a. Water
   b. Urea
   c. Plasma
   d. Calcium

FC *Short answer*

89. List three things that will affect total body water percentage.

90. Name the four fluid compartments.

91. Name four important cations.

92. Identify four ways water normally leaves the body.

CHAPTER 24

Male Reproductive System

HOW TO APPROACH THE MALE REPRODUCTIVE SYSTEM

The male reproductive chapter is very straightforward. Begin with being able to trace sperm from where it is produced to leaving the penis (see the study tip below). Name the glands and what each produces and roughly what percentage they contribute to semen. Make a list of all the hormones produced by or that influence the male reproductive system.

Fill in the blanks with the correct answers.

1. The perineum is divided into the ____________________ triangle and the ____________________ triangle.

2. ____________________ is the production of spermatozoa.

FC 3. Follicle-stimulating hormone (FSH) stimulates the ____________________ tubules to produce ____________________ and luteinizing hormone (LH) stimulates ____________________ cells to secrete ____________________.

4. The ____________________ secrete an alkaline, viscous, creamy-yellow liquid that makes up the majority of semen.

FC 5. The ____________________ creates a pouch for each testis and can dramatically elevate the testes when needed.

6. The ____________________ is composed of three cylindrical structures held together by a covering of skin.

FC *Matching – match each description, word, or phrase with the correct term.*

	Term		Definition
7. ______	Interstitial cells	a.	Paired cylindrical erectile tissue
8. ______	Seminal vesicles	b.	Produces mucus for lubrication
9. ______	Gametes	c.	"Make testosterone"
10. ______	Epididymis	d.	"Make gametes"
11. ______	Prostate	e.	Head of penis
12. ______	Gonads	f.	"Sperm and eggs"
13. ______	Bulbourethral glands	g.	"Make sperm"
14. ______	Corpora cavernosa	h.	Produces fructose
15. ______	Corpus spongiosum	i.	Stores, nourishes, and matures sperm
16. ______	Seminiferous tubules	j.	Covers glans penis
17. ______	Glans penis	k.	Can cause difficulty urinating
18. ______	Prepuce	l.	Contains the urethra

Multiple choice – select the best answer.

19. The area between the thighs shaped like a diamond is called the:
   a. groin.
   b. perineum.
   c. pelvic triangle.
   d. inferior pelvic triangle.

FC 20. The testes are suspended in the scrotum by which of the following?
   a. Seminiferous tubules
   b. Spermatic ligament
   c. Spermatic cords
   d. Spermatic tendon

FC 21. All of the seminiferous tubules come together to form a network called the:
   a. rete testis.
   b. efferent ductules.
   c. spermatic cord.
   d. tunica albuginea.

FC 22. The dense, white, fibrous capsule that covers each testis and divides it into lobules is called the tunica:
   a. vaginalis.
   b. testis.
   c. externa.
   d. albuginea.

23. Which of the following is NOT an effect of testosterone?
   a. Increases length of bones
   b. Stimulates protein anabolism
   c. Promotes hair growth
   d. Closure of epiphyses in long bones

24. What is the name of the cap containing enzymes on the head of the sperm?
   a. Head piece
   b. Spermatic cap
   c. Acrosome
   d. Cephalic layer

25. Which of the following is true regarding the ejaculatory ducts?
   a. Pass through the prostate
   b. Connect testis to seminal vesicle
   c. Are located anterior to the bladder
   d. Connect the bladder to the prostate

26. Which of the following is the correct sequence of glands as sperm leaves the body?
   a. Bulbourethral, prostate, seminal vesicle
   b. Seminal vesicle, prostate, bulbourethral
   c. Prostate, seminal vesicle, bulbourethral
   d. Prostate, bulbourethral, seminal vesicle

27. Which of the following is the correct path sperm takes leaving the body?
    a. Seminiferous tubules, rete testis, efferent ductules, epididymis, vas deferens
    b. Efferent ductules, seminiferous tubules, epididymis, rete testis, ductus deferens
    c. Rete testis, seminiferous tubules, vas deferens, epididymis, efferent ductules
    d. Seminiferous tubules, efferent ductules, rete testis, epididymis, vas deferens

FC 28. What is the name of the membrane that lines each pouch that the testis sits in?
    a. Tunica vaginalis
    b. Tunica interna
    c. Scrotal serosa
    d. Scrotal tunica

29. The glans penis is actually part of what structure?
    a. Spermatic cord
    b. Corpora cavernosa
    c. Prepuce
    d. Corpus spongiosum

FC 30. The creation of sperm is called:
    a. spermatozation.
    b. spermatogenesis.
    c. gametogenesis.
    d. follicularogenesis.

31. What is located superficial to the glans penis?
    a. Corpora cavernosa
    b. Prepuce
    c. Corpus spongiosum
    d. Tunica externa

FC 32. The combination of fluid from the epididymis and the glands is called:
    a. ejaculatory fluid.
    b. testicular semen.
    c. semen.
    d. ejaculatory semen.

Labeling – label the following diagrams.

33. Male reproductive organs

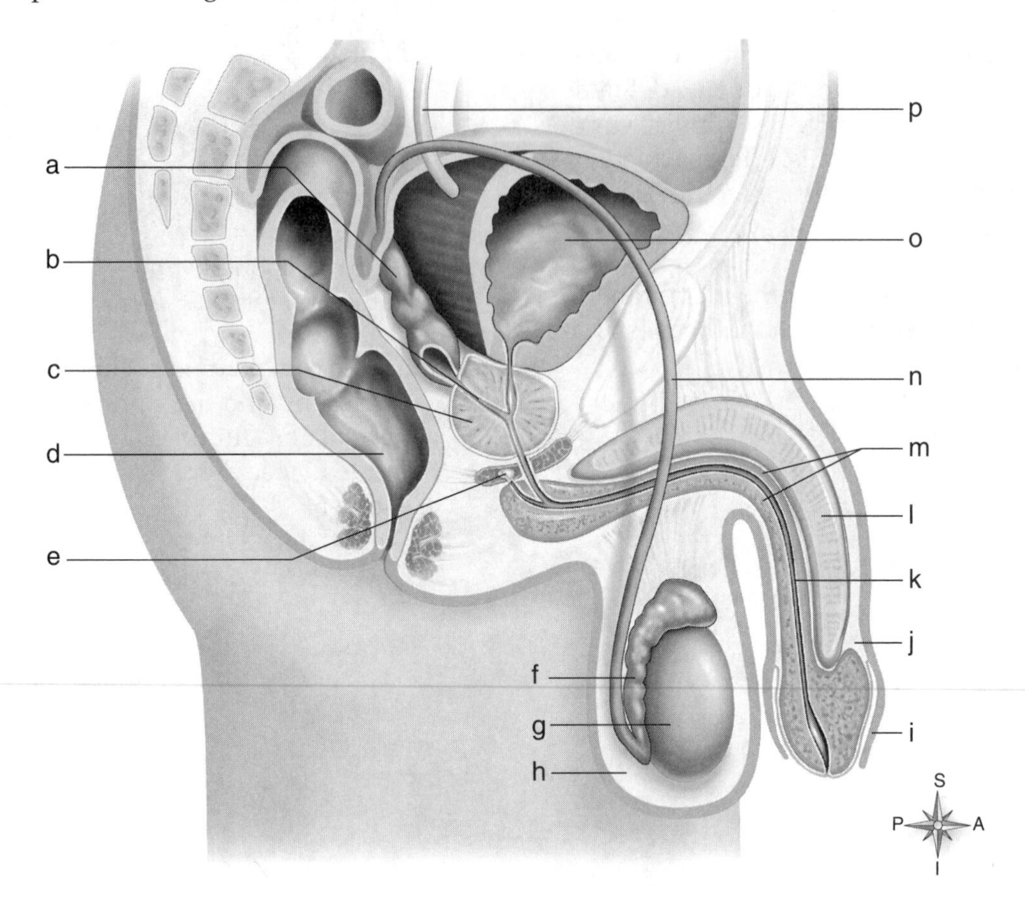

a. ____________________

b. ____________________

c. ____________________

d. ____________________

e. ____________________

f. ____________________

g. ____________________

h. ____________________

i. ____________________

j. ____________________

k. ____________________

l. ____________________

m. ____________________

n. ____________________

o. ____________________

p. ____________________

34. The male reproductive system

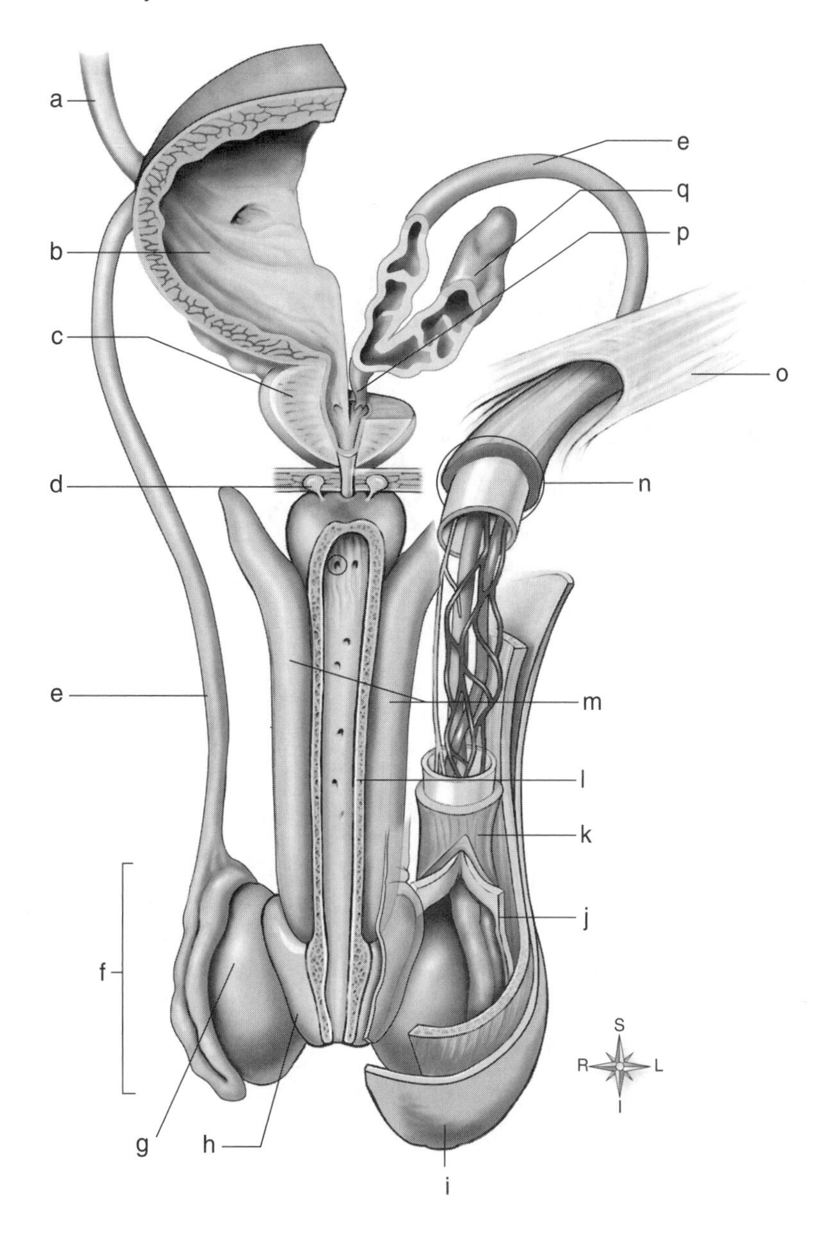

a. ______________________ j. ______________________

b. ______________________ k. ______________________

c. ______________________ l. ______________________

d. ______________________ m. ______________________

e. ______________________ n. ______________________

f. ______________________ o. ______________________

g. ______________________ p. ______________________

h. ______________________ q. ______________________

i. ______________________

35. Testis and epididymis

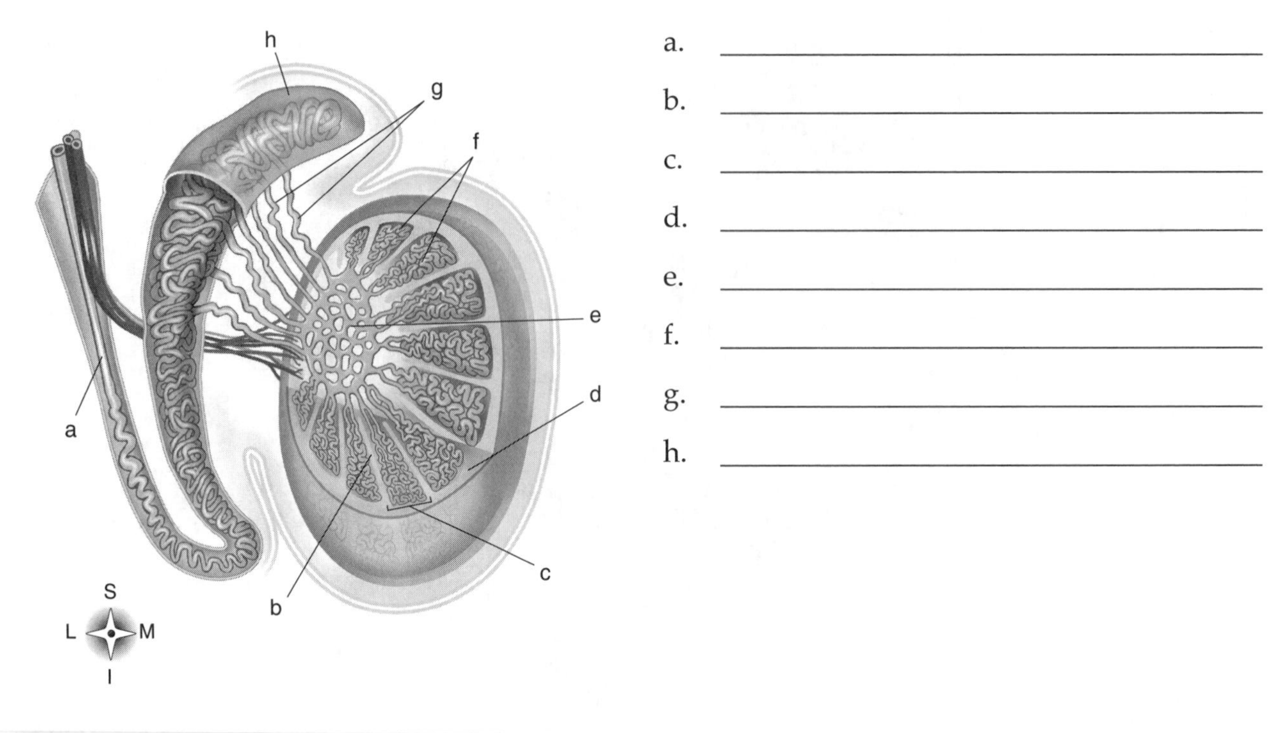

STUDY TIP

After you have learned the female anatomy, trace sperm from production to fertilization in the infundibulum in the female. Using the same method as in Chapter 17, list everything that you pass by or through on a scratch piece of paper. You can test yourself on both chapters in one quick exercise. This will test you on a majority of the structures. When you can do this, then you can move to the next task.

FC *Short answer*

36. Name three things found in the spermatic cords.

37. List the two primary functions of the testes.

38. Name four parts/structures of a sperm.

39. Name the three accessory reproductive glands.

40. List four structures of the penis.

CHAPTER 25

Female Reproductive System

HOW TO APPROACH THE FEMALE REPRODUCTIVE SYSTEM

The female reproductive system is much more complicated than that of the male. I would first approach the chapter with learning the anatomy or the "plumbing." As stated in the previous chapter, just follow the path of sperm to the female infundibulum. Once you have the anatomy down, it's time to focus on the hormones, ovum (egg) development, and the cycles.

First of all, most of the hormones have previously been covered but now you know the actual anatomy they affect, so it is almost like learning them over. You may want to make a list or chart showing the hormone, target, and effect. Keep it simple! Estrogen builds up the endometrium and progesterone maintains it. Follicle-stimulating hormone (FSH) brings a woman to ovulation whereas luteinizing hormone (LH) causes ovulation. It is easy to get lost in the effects of hormones, so break it down and keep it simple.

Getting comfortable with the hormones will lead you into egg development and the cycles. Make sure you understand what a follicle is and how and why it matures. It all has to do with the hormones and their effects on the follicle and the endometrium. Think of the egg as a seed and the endometrium as the garden. Seeds need to be grown, matured, and released. You can't just throw seeds on the ground and expect much to grow. The garden has to be worked and tilled so that the dirt is soft, loose, and thick so that a seed has the best chance of growing. Does this make sense? Just remember to keep it simple; this may mean making some of your own charts, diagrams, or notes.

Fill in the blanks with the correct answers.

1. ____________________ cells completely surround the primary follicle and begin secreting increasing amounts of estrogen-rich fluid that pools around the ____________________ in an enlarging space called the ____________________.

FC 2. The ____________________ follicle matures into a secondary follicle and eventually a(n) ____________________ follicle.

FC 3. The release of an egg from a mature follicle is called ____________________.

4. The ____________________ are really extensions of the uterus that communicate loosely with the ovaries.

FC 5. The ____________________ functions as a temporary endocrine gland secreting progesterone and estrogen.

FC 6. ____________________ stimulates production of milk and ____________________ stimulates ejection of milk into the ducts.

FC ***Matching – match each description, word, or phrase with the correct term.***

	Term		Definition
7. ______	Corpus luteum	a.	Develops the oocyte (egg)
8. ______	Menopause	b.	"Last cycle"
9. ______	Estrogens	c.	Divided into four major phases
10. ______	Ovarian cycle	d.	Maintains the endometrium
11. ______	FSH	e.	Builds up the endometrium
12. ______	LH	f.	Produces mainly progesterone
13. ______	Progesterone	g.	Luteal and follicular phase
14. ______	Menarche	h.	"The ovulating hormone"
15. ______	Endometrial (menstrual) cycle	i.	"First cycle"

FC ***Matching – match each description, word, or phrase with the correct term.***

	Term		Definition
16. ______	Fimbriae	a.	Where the uterus dips into the vagina
17. ______	Cervix	b.	Erectile tissue
18. ______	Endometrium	c.	External genitalia
19. ______	Infundibulum	d.	Varies in thickness throughout the month
20. ______	Hymen	e.	Funnel structure at end of uterine tube
21. ______	Fornix	f.	Space between the labia minora
22. ______	Vulva	g.	Semen often pools here
23. ______	Clitoris	h.	Directly lateral to vaginal orifice
24. ______	Labia minora	i.	Fingerlike extensions of the uterine tube
25. ______	Vestibule	j.	Forms a border around the vaginal orifice

Multiple choice – select the best answer.

26. What anchors the ovary to the uterus?
    a. Uterine ligament
    b. Round ligament
    c. Ovarian ligament
    d. Broad ligament

FC 27. What is the name of the structure that contains an oocyte and specialized cells surrounding it?
    a. Ovarian follicle
    b. Granulose
    c. Ovum
    d. Egg

FC 28. The lower, narrow part of the uterus is the:
    a. fundus.
    b. fornix.
    c. cervix.
    d. infundibulum.

29. What is the fetal organ that permits exchange of materials between the offspring's blood and the maternal blood?
    a. Uterus
    b. Placenta
    c. Endometrium
    d. Umbilical vessels

30. Sperm and ova most often meet and fertilize where?
    a. Ampulla of uterine tube
    b. Vagina
    c. Body of uterus
    d. Cervix of uterus

31. Which of the following is directly medial to the labia majora?
    a. Labia minora
    b. Clitoris
    c. Vestibule
    d. Vaginal orifice

32. Which of the following is homologous with the bulbourethral glands and produces mucus for lubrication?
    a. Glans clitoris
    b. Greater vestibular glands
    c. Vaginal glands
    d. Vaginalmucosal gland

33. Which of the following is NOT located in the vestibule?
    a. Clitoris
    b. Vaginal orifice
    c. External urethral meatus
    d. Labia minora

FC 34. If implantation does not occur, the corpus luteum will reduce to a white scar called the:
    a. follicular scar.
    b. ovarian scar.
    c. corpus albicans.
    d. secondary corpus luteum.

FC 35. Which of the following hormones is a gonadotropin?
    a. LH
    b. Progesterone
    c. Estradiol
    d. Estrone

FC 36. Which of the following causes follicular cells to secrete estrogens?
    a. Progesterone
    b. Estrone
    c. Estradiol
    d. FSH

37. Which of the following is inferior and slightly posterior to the external urethral meatus?
    a. Clitoris
    b. Labia minora
    c. Vaginal orifice
    d. Vestibule

FC 38. Which of the following phases is dominated by high levels of progesterone?
  a. Luteal phase
  b. Follicular phase
  c. Ovarian phase
  d. Proliferative phase

FC 39. Which of the following phases has a rapid rise and high levels of estrogen?
  a. Follicular phase
  b. Luteal phase
  c. Secretory phase
  d. Premenstrual phase

40. Which of the following is positioned so as to be squeezed by a suckling baby?
  a. Mammary glands
  b. Ductules
  c. Lactiferous ducts
  d. Lactiferous sinuses

FC 41. Which area of the breast has numerous sebaceous glands?
  a. Nipple
  b. Areola
  c. Lobes
  d. Lobules

42. All of the following are benefits of breast-feeding EXCEPT:
  a. excellent source of nutrients.
  b. provides active immunity.
  c. enhances emotional bond.
  d. provides exact proportions of nutrients.

Labeling – label the following diagrams.

43. Female reproductive organs

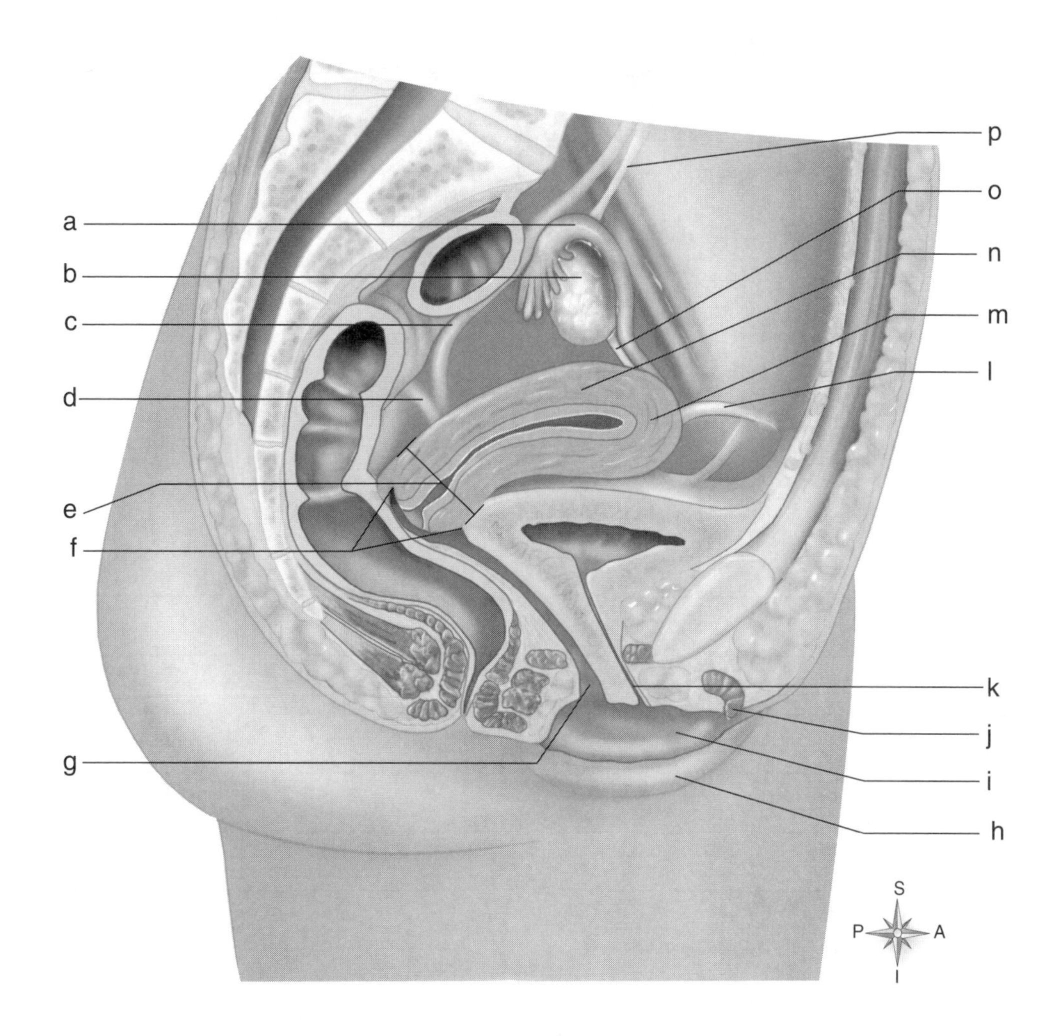

a. ______________________ i. ______________________

b. ______________________ j. ______________________

c. ______________________ k. ______________________

d. ______________________ l. ______________________

e. ______________________ m. ______________________

f. ______________________ n. ______________________

g. ______________________ o. ______________________

h. ______________________ p. ______________________

44. Internal female reproductive organs

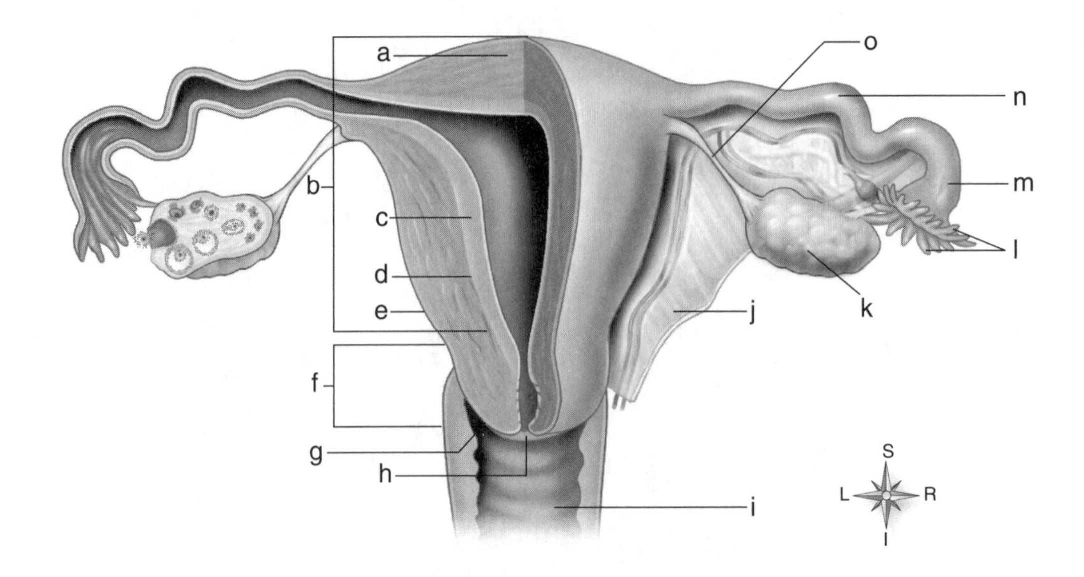

a. ______________________

b. ______________________

c. ______________________

d. ______________________

e. ______________________

f. ______________________

g. ______________________

h. ______________________

i. ______________________

j. ______________________

k. ______________________

l. ______________________

m. ______________________

n. ______________________

o. ______________________

45. Stages of ovarian follicle development

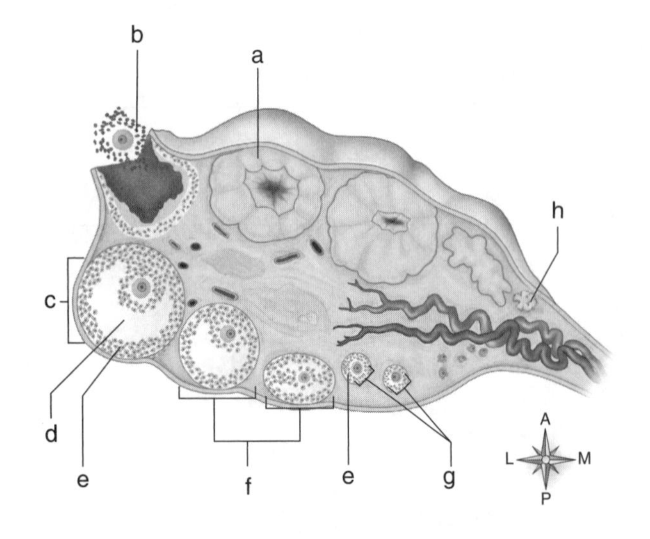

a. ______________________

b. ______________________

c. ______________________

d. ______________________

e. ______________________

f. ______________________

g. ______________________

h. ______________________

46. Vulva

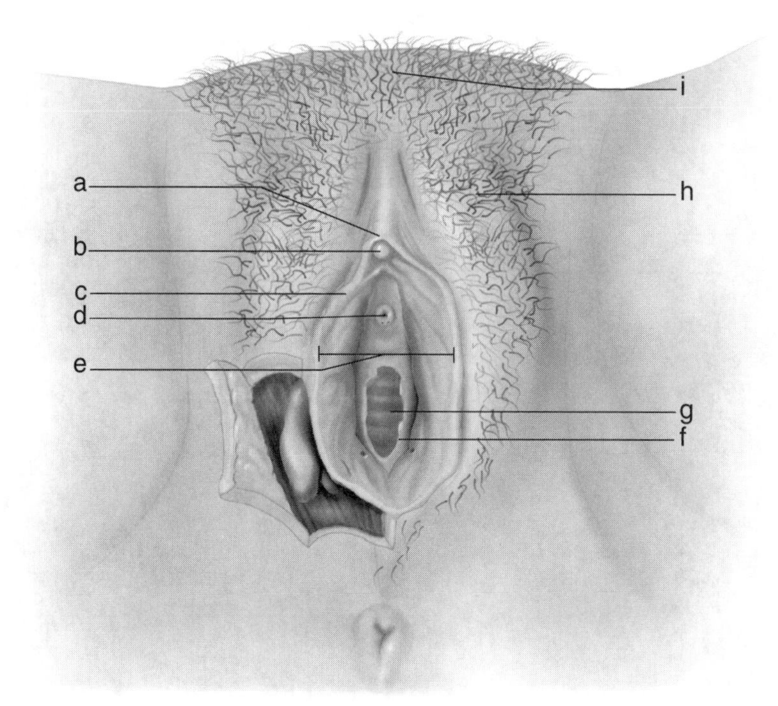

a. ______________________

b. ______________________

c. ______________________

d. ______________________

e. ______________________

f. ______________________

g. ______________________

h. ______________________

i. ______________________

47. The female breast

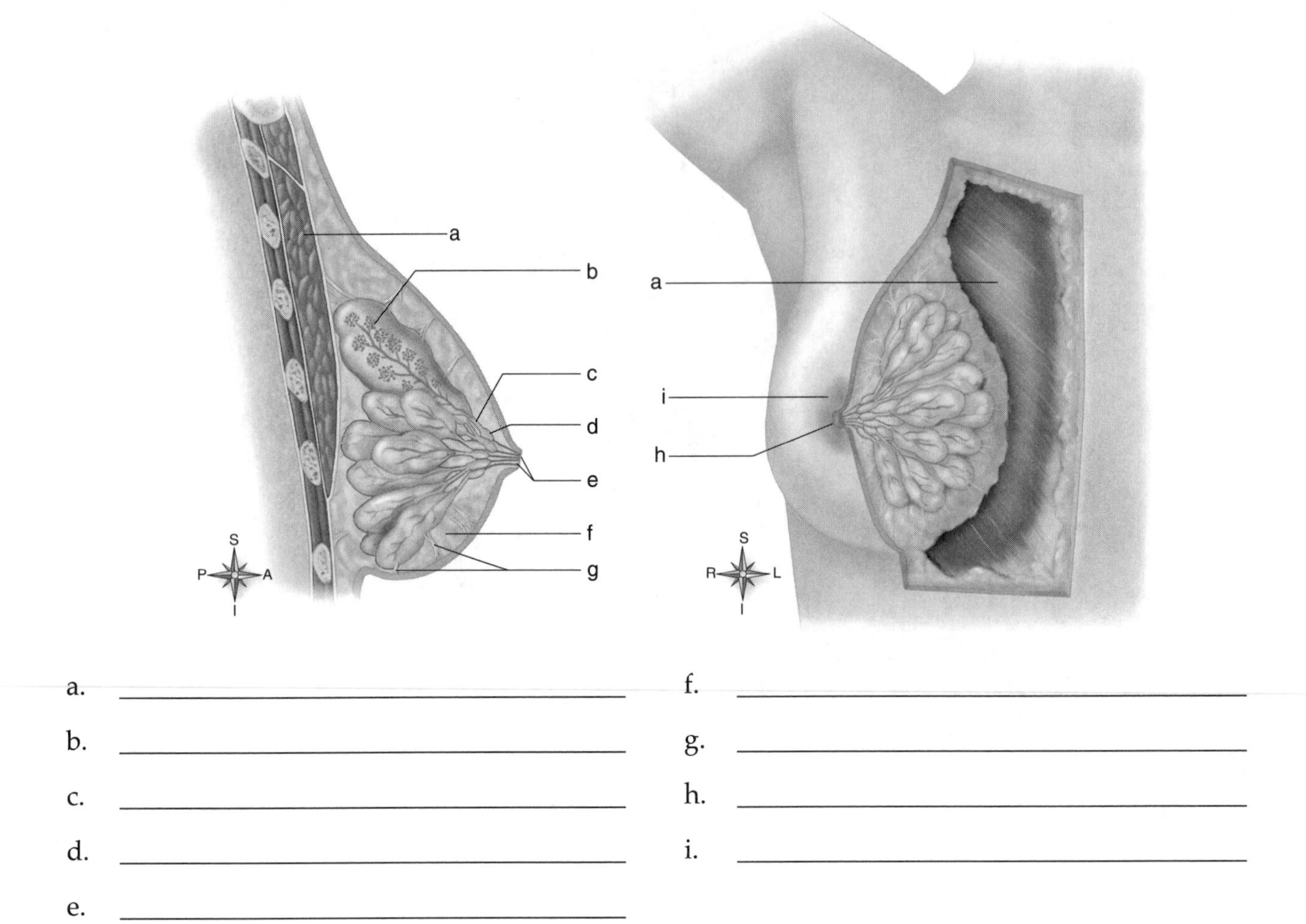

a. ______________________

b. ______________________

c. ______________________

d. ______________________

e. ______________________

f. ______________________

g. ______________________

h. ______________________

i. ______________________

FC *Short answer*

48. What are the two phases of the ovarian cycle?

49. Name five different uterine ligaments.

50. Name the three major sections of the uterus.

51. Name the layers of the uterus superficial to profundus.

52. List the four phases of the endometrial or menstrual cycle.

53. List four effects of estrogen on the uterus.

54. List three effects of progesterone on the uterus.

CHAPTER 26

Growth and Development

HOW TO APPROACH GROWTH AND DEVELOPMENT

It is very important that you understand the previous two chapters and have a very good grasp of the hormones involved in the reproductive system. This chapter takes what you have leaned from the other chapters and puts it all together to explain the process of meiosis, fertilization, and development of a fetus. Take your time with meiosis and make sure you understand what is happening before moving on. You will need to refer to the diagrams to help with this. A good picture is worth a thousand words. If you don't like the pictures/diagrams provided, draw your own.

If you understand meiosis, the rest will fall into place. It makes complete sense how two haploid cells were created the way they were, come together to create our children, and why many times they look nothing like us. If you are not finding what you need to explain something, you may need to go to the companion Evolve website to the *A&P Connect* section.

Fill in the blanks with the correct answers.

FC 1. Production of sperm is called ____________________ and production of ova is called ____________________.

2. ____________________ is the reduction and distribution of chromosomes to make gametes.

FC 3. When paired chromosomes join together to form ____________________, frequently genetic material is exchanged between the paired chromosomes in a process called ____________________.

4. ____________________ cells develop around each primary oocyte, forming a primary follicle.

FC 5. The hollow ball of cells with a fluid-filled center that forms after fertilization is called a(n) ____________________.

FC 6. ____________________ are unspecialized cells that reproduce to form specialized cells.

FC 7. At the end of pregnancy, ____________________ levels of cortisol trigger a drop in ____________________, which in turn causes a drop in progesterone levels. Progesterone inhibits the release of ____________________, which keeps the uterine muscle from contracting. Oxytocin is then released in increasing amounts in a positive feedback mechanism that ____________________ the rate and strength of contractions until delivery.

STUDY TIP

Human chorionic gonadotropin (hCG) is a very important hormone in pregnancy. Unlike the other hormones, we have not really discussed this hormone, so make a flash card with all the important information.

FC *Matching – match each description, word, or phrase with the correct term.*

	Term		Definition
8. ______	Meiosis	a.	23 total chromosomes
9. ______	Diploid cells	b.	Paired chromosomes joined together resulting in four chromatids
10. ______	Spermatogonia	c.	"Suspended animation" until puberty
11. ______	Tetrads	d.	Maturing follicle in response to estrogen
12. ______	Primary follicle	e.	Can be seen as fluid-filled bump on surface of ovary
13. ______	Tertiary follicle	f.	Egg-forming cells
14. ______	Haploid cells	g.	23 pairs of chromosomes
15. ______	Secondary follicle	h.	Sperm-forming cells
16. ______	Oogonia	i.	"Reshuffling" genes
17. ______	"Crossing over"	j.	Creation of sperm and eggs

FC *Matching – match each description, word, or phrase with the correct term.*

	Term		Definition
18. ______	Zygote	a.	Anchors fetus to uterus
19. ______	Blastocyst	b.	Outer wall of blastocyst
20. ______	Trophoblast	c.	"Bag of water"
21. ______	Morula	d.	Stage after morula
22. ______	Implantation	e.	Blood vessel extensions of the chorion
23. ______	Embryology	f.	Cell right after fertilization
24. ______	Chorion	g.	Study of prenatal period
25. ______	Chorionic villi	h.	Embedding in the endometrium
26. ______	Placenta	i.	Solid mass of cells about 3 days after fertilization
27. ______	Amniotic cavity	j.	Important fetal membrane in the placenta

STUDY TIP

The primary germ layers are consistent with the names of some of the membranes or layers you have learned from previous chapters. The name tells you what they are going to form or where they would be located. ENDOderm forms the linings just like endocardium and endometrium. MESOderm is similar to myocardium or myometrium and ECTOderm refers to outside or covering.

Multiple choice – select the best answer.

FC 28. Meiosis will create which of the following?
- a. Diploid cells
- b. Triploid cells
- c. Haploid cells
- d. Monoloid cells

FC 29. Which of the following has 23 pairs of chromosomes?
- a. Diploid cells
- b. Spermatogonia
- c. Oogonia
- d. Haploid cells

30. Chromatids are connected by a single:
- a. centromere.
- b. centriole.
- c. spindle fiber.
- d. tetrad.

31. How many total cells are produced from both meiosis I and meiosis II?
- a. 2
- b. 3
- c. 4
- d. 8

32. A mature follicle ready to burst open is frequently called a:
- a. secondary follicle.
- b. graafian follicle.
- c. polar body.
- d. zygote.

33. During oogenesis, the cytoplasm is mostly given to only one of the haploid daughter cells. What is the name of the remaining three cells?
- a. Ovum
- b. Polar bodies
- c. Graafian cells
- d. Folliculocytes

34. What is the end result of spermatogenesis?
- a. One functional haploid cell and the polar bodies
- b. Two functional haploid cells and two polar bodies
- c. Three functional haploid cells and one polar body
- d. Four mature sperm

35. Which of the following does NOT assist sperm movement in the female?
- a. Mucus strands in the cervical canal
- b. Release of progesterone with orgasm
- c. Rhythmic contractions of the female reproductive tract
- d. Ciliary movement in uterine tubes

FC 36. A fertilized ovum is called a(n):
- a. egg.
- b. zygote.
- c. blastocyst.
- d. morula.

FC 37. Which of the following implants in the uterine lining?
- a. Morula
- b. Zygote
- c. Embryo
- d. Blastocyst

38. What does the uterine lining secrete after fertilization?
    a. hCG
    b. Uterine milk or histotrophe
    c. Progesterone only
    d. Progesterone and estrogen

39. The yolk sac and amniotic cavity develop from:
    a. inner cell mass.
    b. trophoblast.
    c. chorion.
    d. placenta.

40. Which of the following develops from the trophoblast?
    a. Yolk sac
    b. Amniotic cavity
    c. Placenta
    d. Chorion

FC 41. What secretes large amounts of hCG?
    a. Placenta
    b. Endometrium
    c. Ovary
    d. Corpus luteum

FC 42. hCG stimulates the:
    a. endometrium to release inhibin.
    b. corpus luteum to continue secreting estrogen and progesterone.
    c. pituitary gland to release follicle-stimulating hormone (FSH) and luteinizing hormone (LH).
    d. ovary to halt follicle development.

FC 43. What is tested for in pregnancy tests?
    a. Levels of FSH and LH
    b. Increasing levels of estrogens
    c. Increasing levels of estrogens and hCG
    d. hCG

FC 44. For the majority of the gestation period, the "baby" is called a(n):
    a. embryo.
    b. fetus.
    c. blastocyst.
    d. zygote.

45. Adult stem cells are called:
    a. totipotent cells.
    b. pluripotent cells.
    c. multipotent stem cells.
    d. adult zygote cells.

FC 46. What forms the many structures around the periphery of the body?
    a. Ectoderm
    b. Epiderm
    c. Endoderm
    d. Myoderm

FC 47. The lining of the respiratory and digestive tracts are formed by which germ layer?
    a. Ectoderm
    b. Epiderm
    c. Mesoderm
    d. Endoderm

FC 48. Most of the organs and skeletal muscles are formed by which germ layer?
    a. Myoderm
    b. Endoderm
    c. Mesoderm
    d. Ectoderm

49. The process by which the primary germ layers develop into different kinds of tissue is called:
    a. dermogenesis.
    b. germogenesis.
    c. histogenesis.
    d. organogenesis.

50. To assess the general condition of a newborn, a system that scores five health criteria is called the:
    a. neonatology assessment.
    b. neonatal score.
    c. embryonic delivery score.
    d. Apgar score.

51. The study of aging is called:
    a. gerontology.
    b. senescence.
    c. ageology.
    d. parturition.

FC *Short answer*

52. Identify the three primary germ layers.

53. Name three structures that form the ectoderm.

54. Name three structures that form the endoderm.

55. Name three structures that form the mesoderm.

56. List the three stages of labor and give a BRIEF description of each.

57. Identify the four postnatal periods of life.

CHAPTER 27

Human Genetics and Heredity

HOW TO APPROACH GENETICS AND HEREDITY

Genetics and heredity build on previous chapters. You may need to go back and review meiosis and cell reproduction. It's always important, especially in this chapter, to make sure you have memorized and understand the terms *gene, chromosome, genome,* etc., before moving on. I might go as far as making and reviewing some flash cards over the basic terms before finishing the chapter. At the very least, make sure you understand them before moving on.

It's all about "genes." The bulk of this chapter revolves around genes. Their locations, replication, presentation, and function are what heredity is all about. Once you understand these things, using a Punnett square makes complete sense. You also now understand the importance of a "pedigree" when purchasing a dog or racing horse.

Fill in the blanks with the correct answers.

FC 1. ____________________ is the scientific study of inheritance.

FC 2. DNA in its normal state and not replicated is called ____________________. When the DNA is fully condensed and replicated, it is called a(n) ____________________.

FC 3. The entire collection of genetic material in a typical cell is called the ____________________.

FC 4. A(n) ____________________ gene is one whose effects are seen and are capable of masking the effects of a(n) ____________________ gene for the same trait.

5. Inherited characteristics that are determined by the combined effect of many different gene pairs are often called *polygenic traits* to distinguish them from ____________________.

6. A form of dominance called ____________________ will have roughly equal effects of the genes, displaying traits from both.

7. A(n) ____________________ is a change in an individual's genetic code.

8. The shorter segment of the chromosome is called the ____________________ and the longer segment is called the ____________________.

FC 9. The male chromosome is the "Y" chromosome and the female chromosome is the __________ chromosome.

10. Fetal tissue can be collected by a(n) ____________________ or ____________________ to screen for genetic abnormalities.

FC *Matching – match each description, word, or phrase with the correct term.*

	Term		Definition
11. ______	Chromosome	a.	Basic genetic "unit"
12. ______	Histones	b.	Subunits of chromatin
13. ______	Genes	c.	"All" your genetic material
14. ______	Genome	d.	The combination of your genes like AA or Aa
15. ______	Genotype	e.	The manner in which the genotype is expressed
16. ______	Homozygous	f.	Condensed and paired chromatin
17. ______	Phenotype	g.	Identical genotypes such as AA or aa
18. ______	Heterozygous	h.	Single thread of DNA wound around histones
19. ______	Nucleosomes	i.	Different genotypes such as Aa
20. ______	Chromatin	j.	Protein molecules found in strands of chromatin

Multiple choice – select the best answer.

FC 21. A sequence of nucleotide bases in the DNA molecule is called a:
 a. gene.
 b. genome.
 c. genotype.
 d. chromatid.

FC 22. What is seen on TV and the movies that is shaped like an "X" and refers to our genetic material?
 a. Chromatin
 b. Gene
 c. DNA
 d. Chromosome

FC 23. Which of the following contains a "recipe" for one specific polypeptide?
 a. Chromosome
 b. Chromatin
 c. Gene
 d. Genome

FC 24. A double-helix strand of genetic material that looks like a ladder without other structures such as proteins is called:
 a. chromosome.
 b. DNA.
 c. chromatin.
 d. genes.

25. Each sperm and egg are likely to receive a different set of 23 chromosomes due to the principle of:
 a. independence.
 b. independent assortment.
 c. independent segregation.
 d. differentiation.

26. According to what principle, genes on an individual chromosome tend to stay together?
    a. Principle of gene linkage
    b. Principle of adhesiveness
    c. Principle of cohesion
    d. Principle of simplicity

27. In an example of a genotype Aa, the A is what?
    a. Polygenic
    b. The dominant gene
    c. The recessive gene
    d. The phenotype

FC 28. Which of the following would you call a genotype of AA?
    a. Homozygous dominant
    b. Heterozygous dominant
    c. Homozygous recessive
    d. Heterozygous recessive

29. Which of the following would you call a genotype of Aa?
    a. Homozygous
    b. Heterozygous
    c. Polygenic
    d. Monogenic

30. A mutation in which one or more nucleotide bases in a sequence is missing is called a(n):
    a. insertion.
    b. regression.
    c. deletion.
    d. subtraction.

FC 31. Expressed genotypes are called:
    a. phenotypes.
    b. carriers.
    c. genetic presentations.
    d. genomes.

32. Characteristics such as skin color, which are determined by the combined effect of several different gene pairs, are called:
    a. multigenic.
    b. polygenic.
    c. digenic.
    d. monogenic.

33. Traits carried on sex chromosome are called:
    a. XY traits.
    b. sex traits.
    c. reproductive-linked traits.
    d. sex-linked traits.

34. Which of the following is correct related to X-linked recessive traits?
    a. Most likely to cause mutations
    b. Appear much more frequently in males than in females
    c. Appear much more frequently in females than in males
    d. Are frequently call mutagens

35. Which of the following is/are NOT considered a mutagen?
    a. Bacteria
    b. Chemicals
    c. Radiation
    d. Viruses

36. Which of the following is a chart illustrating genetic relationships in a family over several generations?
    a. Punnett square
    b. Pedigree
    c. Genetic graph
    d. Case chart

37. Which of the following is a grid used to determine the mathematical probability of inheriting genetic traits?
    a. Punnett square
    b. Pedigree
    c. Genetic graph
    d. Case chart

Answer Key

CHAPTER 1: ORGANIZATION OF THE HUMAN BODY

CHARACTERISTICS OF LIFE AND LEVELS OF ORGANIZATION

Fill in the Blanks

1. molecules, macromolecules
2. organelles
3. cells
4. tissues, organs, organ systems, organism

Matching

5. b
6. e
7. h
8. a
9. j
10. d
11. c
12. f
13. i
14. g

Multiple Choice

15. b
16. b
17. d
18. d
19. b

BODY CAVITIES AND REGIONS

Fill in the Blanks

20. dorsal, cranial
21. thoracic, abdominopelvic
22. thoracic, pleural, mediastinum
23. parietal, visceral
24. serous fluid

Labeling

25. Major body cavities
    a. Cranial cavity
    b. Spinal cavity
    c. Thoracic cavity
    d. Pleural cavities
    e. Mediastinum
    f. Abdominal cavity
    g. Pelvic cavity
    h. Dorsal
    i. Ventral
26. Nine regions of the abdominopelvic cavity
    a. Right hypochondriac
    b. Epigastric
    c. Left hypochondriac
    d. Right lumbar
    e. Umbilical region
    f. Left lumbar
    g. Right iliac
    h. Hypogastric
    i. Left iliac

Multiple Choice

27. d
28. d
29. c
30. c
31. a

ANATOMICAL TERMS, BODY PLANES AND SECTIONS

Matching

32. j
33. i
34. a
35. b

36. c
37. d
38. e
39. f
40. g
41. h
42. h
43. j
44. g
45. b
46. i
47. c
48. e
49. f
50. a
51. d

Multiple Choice
52. b
53. a
54. a
55. c
56. d
57. b
58. c
59. b
60. c
61. c

Labeling
62. Specific body regions, ventral view
    a. Cephalic
    b. Axillary
    c. Mammary
    d. Brachial
    e. Antecubital
    f. Antebrachial
    g. Carpal
    h. Digital or phalangeal
    i. Femoral
    j. Crural
    k. Tarsal
    l. Digital
    m. Pedal
    n. Patellar
    o. Inguinal
    p. Palmar
    q. Abdominal
    r. Thoracic
    s. Zygomatic
    t. Orbital
    u. Frontal
63. Specific body regions, dorsal view
    a. Occipital
    b. Cervical
    c. Brachial
    d. Olecranal
    e. Antebrachial
    f. Gluteal
    g. Femoral
    h. Popliteal
    i. Calcaneal
64. Directions and planes of the body
    a. Superior
    b. Posterior
    c. Anterior
    d. Proximal
    e. Distal
    f. Inferior
    g. Transverse plane
    h. Lateral
    i. Frontal plane
    j. Sagittal plane
    k. Medial

HOMEOSTASIS

Fill in the Blanks
65. homeostasis
66. homeostatic control mechanisms, feedback control loop
67. sensor, integrator, effector, feedback (any order)
68. afferent, efferent
69. negative, positive
70. opposite, stabilizes
71. amplifies or accelerates

CHAPTER 2: THE CHEMISTRY OF LIFE

BASIC CHEMISTRY

Fill in the Blanks
1. synthesis
2. chemical
3. decomposition
4. exchange
5. metabolism

Matching
6. h
7. f
8. j
9. a
10. b
11. c
12. d
13. e
14. g
15. i

Multiple Choice
16. b
17. c
18. c
19. a
20. c
21. b
22. c
23. d
24. d
25. c
26. a
27. b
28. a
29. b
30. d

Short Answer
31. Synthesis, decomposition and exchange
32. Catabolism and anabolism
33. Solid, liquid, gas
34. Protons, neutrons, electrons
35. Ionic, covalent, hydrogen

INORGANIC MOLECULES

Multiple Choice
36. c
37. b
38. a
39. c
40. d
41. d
42. b
43. c
44. a

ORGANIC MOLECULES

Matching
45. j
46. g
47. b
48. d
49. i
50. a
51. c
52. f
53. e
54. h

Short Answer
55. DNA, RNA
56. Adenine, thymine, cytosine, guanine
57. Carbon, hydrogen, oxygen, and nitrogen
58. Carbohydrates, proteins, lipids, nucleic acids
59. Triglycerides, phospholipids, steroids, prostaglandins

CHAPTER 3: ANATOMY OF CELLS

THE FUNCTIONAL ANATOMY OF CELLS

Multiple Choice
1. c
2. a
3. c
4. a
5. d
6. b

BASIC ORGANELLES AND THEIR FUNCTIONS

Fill in the Blanks
7. membranous, nonmembranous
8. endoplasmic reticulum, rough ER, smooth ER
9. peroxisomes

Matching
10. d
11. g
12. e
13. i
14. b

15. a
16. h
17. f
18. c

Multiple Choice
19. d
20. b
21. c
22. c
23. a
24. a

NUCLEUS, CYTOSKELETON, AND CELL CONNECTIONS

Matching
25. b
26. i
27. j
28. a
29. f
30. g
31. c
32. h
33. d
34. e

Multiple Choice
35. c
36. a
37. c
38. a
39. c
40. a
41. b
42. a
43. b
44. c

Labeling
45. Typical, or composite, cell
 a. Nucleolus
 b. Nuclear envelope
 c. Nucleus
 d. Centrioles
 e. Centrosome
 f. Mitochondrion
 g. Golgi apparatus
 h. Free ribosomes
 i. Lysosome
 j. Microvilli
 k. Smooth endoplasmic reticulum
 l. Cilia
 m. Rough endoplasmic reticulum
 n. Chromatin

CHAPTER 4: PHYSIOLOGY OF CELLS

MOVEMENT OF SUBSTANCES THROUGH CELL MEMBRANES

Fill in the Blanks
1. membrane pumps
2. three, two, ATP
3. diffusion, filtration
4. hypotonic, hypertonic

Matching
5. e
6. f
7. b
8. j
9. h
10. c
11. d
12. a
13. i
14. g

Multiple Choice
15. b
16. d
17. a
18. a
19. c
20. b
21. c
22. d
23. b
24. a
25. b

Short Answer
26. Endocytosis and exocytosis
27. Phagocytosis and pinocytosis
28. Isotonic, hypotonic, hypertonic
29. Channel-mediated passive transport and carrier-mediated passive transport

CELL METABOLISM

Fill in the Blanks
30. kinases
31. denaturation
32. cell metabolism
33. proteins, cofactor
34. anabolic, catabolic

Multiple Choice
35. a
36. b
37. c
38. b
39. c
40. a

CELL RESPIRATION

Fill in the Blanks
41. cellular respiration
42. anaerobic, aerobic
43. glucose, cytosol, lactic acid, aerobic, citric acid, electron transport
44. catabolism
45. glycolysis, citric acid cycle, electron transport system

Matching
46. e
47. c
48. b
49. f
50. a
51. d

Multiple Choice
52. b
53. b
54. a
55. b
56. d
57. c

CHAPTER 5: CELL GROWTH AND REPRODUCTION

DNA/RNA

Fill in the Blanks
1. molecular genetics
2. deoxyribonucleic acid
3. bases
4. protein
5. tRNA
6. polyribosome

Matching
7. f
8. i
9. a
10. g
11. b
12. c
13. d
14. e
15. h

Multiple Choice
16. c
17. d
18. a
19. a
20. a
21. a
22. c
23. a
24. d

Short Answer
25. Adenine, guanine, cytosine, and thymine
26. Adenine, guanine, cytosine, uracil
27. mRNA, tRNA

Labeling
28. Protein synthesis
    a. DNA
    b. Bases
    c. Codon
    d. Ribosome
    e. Polypeptide
    f. Translation
    g. tRNA
    h. Amino acids
    i. Peptide bond
    j. Transcription
    k. mRNA

CELL GROWTH AND REPRODUCTION

Fill in the Blanks
29. Mitosis, meiosis
30. Spindle fibers
31. Haploid

32. Zygote
33. 46

Matching
34. d
35. g
36. c
37. h
38. a
39. f
40. b
41. e

Multiple Choice
42. d
43. b
44. b
45. a
46. c
47. a
48. d
49. d
50. b
51. b
52. c
53. a

Short Answer
54. Interphase, prophase, metaphase, anaphase, telophase, and cytokinesis
55. Prophase, metaphase, anaphase, telophase
56. S-phase, G1 and G2

Labeling
57. Events of mitosis
    a. Anaphase
    b. Metaphase
    c. Telophase
    d. Prophase

CHAPTER 6: TISSUES AND THEIR FUNCTIONS

PRINCIPAL TYPES OF TISSUE AND EXTRACELLULAR MATRIX

Fill in the Blanks
1. tissues
2. extracellular matrix
3. histology
4. proteoglycans
5. glycoproteins

Matching
6. b
7. a
8. c
9. a
10. d
11. a
12. b
13. c
14. d
15. a

Short Answer
16. Epithelial, connective, muscle, nervous
17. Ectoderm, endoderm, and mesoderm
18. Collagen and elastin

Labeling
19. Tissue types
    a. Simple columnar, epithelium
    b. Neuron, nervous
    c. Smooth, muscle
    d. Simple cuboidal, epithelium
    e. Reticular, connective
    f. Compact bone, connective
    g. Blood, connective
    h. Cardiac, muscle
    i. Adipose, connective
    j. Skeletal, muscle
    k. Areolar, connective
    l. Hyaline cartilage, connective
    m. Pseudostratified columnar, epithelium
    n. Transitional, epithelium
    o. Irregular dense, connective
    p. Elastic cartilage, connective
    q. Fibrocartilage, connective
    r. Stratified squamous
    s. Cancellous or spongy bone, connective

EPITHELIAL TISSUE

Fill in the Blanks
20. membranous, glandular
21. goblet cells
22. microvilli
23. tubular, alveolar

Matching
24. b

25. c
26. d
27. e
28. a
29. g
30. f
31. b
32. c
33. a

Multiple Choice

34. b
35. a
36. c
37. b
38. a
39. a
40. d
41. b

Short Answer

42. Protection, sensory, secretion, absorption, excretion
43. Squamous, cuboidal, columnar, pseudostratified columnar
44. Simple, stratified, and transitional
45. Holocrine, apocrine, merocrine

Labeling

46. Three types of exocrine glands
    a. Apocrine
    b. Holocrine
    c. Merocrine

CONNECTIVE TISSUE

Fill in the Blanks

47. tendons, ligaments
48. regular, irregular
49. osteocytes, osteoblasts, osteoclasts
50. canaliculi, central canal

Matching

51. f
52. i
53. b
54. g
55. c
56. h
57. d
58. e
59. a

Multiple Choice

60. b
61. a
62. d
63. b
64. a
65. d
66. a
67. d
68. b

Short Answer

69. Collagenous, reticular, elastic
70. Hyaline, fibrocartilage, elastic
71. Lacunae, osteocyte, lamellae, canaliculi, central canal
72. Erythrocytes, leukocytes, thrombocytes

MUSCLE AND NERVOUS TISSUE, TISSUE REPAIR, AND BODY MEMBRANES

Fill in the Blanks

73. muscle
74. parietal, visceral
75. pleura, peritoneum
76. pericardium
77. nervous

Matching

78. g
79. f
80. d
81. i
82. j
83. k
84. a
85. c
86. b
87. h
88. e

Multiple Choice

89. d
90. d
91. c
92. d
93. a
94. d
95. b
96. a

Short Answer

97. Skeletal, smooth, cardiac
98. Cutaneous, serous, mucous
99. Stomach, intestines, blood vessels, organs

Labeling

100. Types of body membranes
    a. Mucous
    b. Cutaneous
    c. Visceral pleura
    d. Parietal pleura
    e. Serous: visceral
    f. Serous: parietal
    g. Visceral peritoneum
    h. Parietal peritoneum

CHAPTER 7: SKIN AND ITS APPENDAGES

STRUCTURE OF THE SKIN

Fill in the Blanks

1. tactile, Merkel
2. calluses
3. dermoepidermal junction
4. hypodermis
5. keratin

Matching

6. e
7. g
8. d
9. c
10. h
11. b
12. i
13. j
14. f
15. a

Multiple Choice

16. d
17. a
18. b
19. c
20. d
21. d
22. a
23. c
24. a
25. d
26. a
27. d
28. c

Labeling

29. Layers of the skin
    a. Epidermis
    b. Stratum corneum
    c. Stratum lucidum
    d. Stratum granulosum
    e. Stratum spinosum
    f. Stratum basale
    g. Dermis
    h. Papillary layer
    i. Reticular layer
    j. Hypodermis

Short Answer

30. Dermis and epidermis
31. Corneum, lucidum, granulosum, spinosum, basale
32. Keratinocytes, melanocytes, dendritic cells

SKIN COLOR AND FUNCTIONS

Fill in the Blanks

33. Melanin
34. UV light
35. Cyanotic

Multiple Choice

36. c
37. a
38. d
39. c
40. a
41. b
42. d

Short Answer

43. Protection, sensation, movement, endocrine, excretion, immunity, temperature regulation
44. Fat-soluble vitamins, sex hormones, corticoid hormones, certain drugs such as nicotine and nitroglycerin
45. Touch, pressure, vibration, pain, and temperature
46. A, D, E, K
47. Uric acid, ammonia, and urea

APPENDAGES OF THE SKIN

Fill in the Blanks

48. Lanugo
49. Vellus
50. Sebaceous, sebum

Matching

51. b
52. c
53. a
54. d
55. c
56. a
57. b
58. a
59. b
60. c

Multiple Choice

61. c
62. a
63. b
64. c
65. d
66. b
67. a
68. d
69. d
70. c
71. c

Labeling

72. Structure of nails
    a. Free edge
    b. Nail body
    c. Lunula
    d. Cuticle
    e. Nail root

Short Answer

73. Palms, soles, lips, nipples, areas of genitalia
74. Sweat, sebaceous, ceruminous
75. Triglycerides, waxes, fatty acids, and cholesterol

CHAPTER 8: SKELETAL TISSUES

TYPES OF BONES

Fill in the Blanks

1. compact, spongy
2. endosteum
3. medullary cavity

Matching

4. c
5. e
6. h
7. g
8. i
9. b
10. f
11. j
12. d
13. a

Multiple Choice

14. c
15. a
16. b
17. c
18. d
19. a

Labeling

20. Long bone
    a. Epiphysis
    b. Diaphysis
    c. Periosteum
    d. Yellow marrow
    e. Endosteum
    f. Medullary cavity
    g. Compact bone
    h. Epiphyseal line
    i. Spongy bone
    j. Articular cartilage

Short Answer

21. Long, short, flat, irregular, and sesamoid
22. Diaphysis, epiphyses, articular cartilage, periosteum, medullary cavity, endosteum

BONE TISSUE AND MICROSCOPIC STRUCTURES OF BONE

Fill in the Blanks

23. extracellular bone matrix
24. calcification
25. chondroitin sulfate, glucosamine
26. osteocytes

Matching

27. i
28. h
29. d
30. f
31. e
32. b
33. a
34. j
35. g
36. c

Multiple Choice

37. c
38. d
39. a
40. c
41. b
42. b
43. b
44. d
45. b
46. b

Labeling

47. Compact and cancellous bone in a long bone
    a. Lamellae
    b. Central canal
    c. Trabeculae
    d. Cancellous (spongy) bone
    e. Compact bone
    f. Osteon
    g. Periosteum
    h. Lacunae

Short Answer

48. Lamellae, lacunae, canaliculi, central canal
49. Osteoblasts, osteoclasts, osteocytes

BONE MARROW, REGULATION, DEVELOPMENT, REMODELING, AND REPAIR AND CARTILAGE

Fill in the Blanks

50. osteoclasts
51. osteoblasts
52. fracture
53. fracture hematoma, granulation, callus
54. avascular

Matching

55. c
56. a
57. b
58. a
59. c
60. a
61. a
62. b
63. b
64. c

Multiple Choice

65. c
66. c
67. b
68. c
69. a
70. a
71. d
72. b
73. a
74. c

Labeling

75. Types of cartilage
    a. Hyaline
    b. Elastic
    c. Fibrocartilage

CHAPTER 9: BONES AND JOINTS

INTRODUCTION

Fill in the Blanks

1. 206
2. axial skeleton, appendicular skeleton
3. axial skeleton, appendicular skeleton

Matching
4. b
5. f
6. j
7. h
8. g
9. c
10. i
11. e
12. d
13. a

SKULL

Fill in the Blanks
14. cranial, facial
15. hyoid bone
16. fontanels
17. vomer
18. paranasal sinuses

Short Answer
19. (1) frontal, (2) parietal, (1) temporal, (1) occipital, (1) sphenoid, (1) ethmoid
20. (2) maxillae, (2) zygomatic, (2) nasal, (1) mandible, (2) lacrimal, (2) palatine, (2) inferior nasal conchae, and (1) vomer
21. Sphenoid, ethmoid, maxillae, and frontal
22. Malleus, incus, stapes
23. Lambdoidal, squamous, and coronal

Multiple Choice
24. c
25. a
26. c
27. c
28. a
29. c
30. b
31. c
32. b
33. a
34. a
35. d
36. b

Labeling
37. Anterior view of the skull
    a. Nasal bone
    b. Ethmoid bone
    c. Infraorbital foramen
    d. Maxilla
    e. Mental foramen
    f. Vomer
    g. Inferior nasal concha
    h. Perpendicular plate of the ethmoid
    i. Optic foramen
    j. Sphenoid bone
    k. Supraorbital foramen
    l. Parietal bone
    m. Frontal bone
38. Skull viewed from the right side
    a. Squamous suture
    b. Lambdoid suture
    c. External acoustic meatus
    d. External occipital protuberance
    e. Mastoid process of temporal bone
    f. Styloid process
    g. Zygomatic process of temporal bone
    h. Temporal process of zygomatic bone
    i. Lacrimal bone
    j. Ethmoid bone
    k. Zygomatic arch
    l. Sphenoid bone
    m. Coronal suture
39. Floor of the cranial cavity viewed from above
    a. Crista galli of ethmoid
    b. Cribriform plate
    c. Superior orbital fissure
    d. Foramen ovale
    e. Foramen lacerum
    f. Foramen spinosum
    g. Internal acoustic meatus
    h. Jugular foramen
    i. Foramen magnum
    j. Petrous part of temporal bone
    k. Sella turcica
    l. Greater wing
    m. Lesser wing
    n. Sphenoid bone
40. Skull viewed from below
    a. Incisive foramen
    b. Zygomatic process of maxilla
    c. Zygomatic arch
    d. Styloid process
    e. Foramen ovale
    f. Mastoid process
    g. Foramen magnum
    h. Occipital condyle
    i. Jugular foramen
    j. Vomer
    k. Lateral pterygoid plate of sphenoid
    l. Medial pterygoid plate of sphenoid

m. Palatine bone
n. Palatine process of maxilla
o. Hard palate

VERTEBRAL COLUMN, STERNUM, AND RIBS

Fill in the Blanks

41. 7, 12, 5
42. sacrum
43. xiphoid process, body, manubrium
44. true ribs
45. dens or odontoid process, axis

Labeling

46. Cervical vertebrae, superior and lateral views
    a. Pedicle
    b. Body
    c. Transverse foramen
    d. Spinous process
    e. Lamina
    f. Superior articular facet
    g. Transverse process
47. Thoracic vertebrae, superior and lateral views
    a. Body
    b. Superior articular facet
    c. Transverse process
    d. Lamina
    e. Spinous process
    f. Pedicle
    g. Superior articular process
    h. Inferior articular facet

THE PECTORAL GIRDLE AND UPPER EXTREMITIES

Fill in the Blanks

48. pectoral girdle
49. capitulum, trochlea
50. 8, 5
51. head, radial tuberosity, styloid process

Multiple Choice

52. b
53. a
54. a
55. d
56. d
57. a
58. b
59. a
60. d
61. c

Matching

62. d
63. c
64. e
65. f
66. h
67. g
68. i
69. b
70. j
71. a

Labeling

72. Right scapula, anterior view
    a. Coracoid process
    b. Acromion
    c. Glenoid cavity
    d. Subscapular fossa
    e. Lateral border
    f. Inferior angle
    g. Medial border
    h. Superior angle
    i. Superior border
73. Right scapula, posterior view
    a. Supraspinous fossa
    b. Infraspinous fossa
    c. Glenoid cavity
    d. Acromion
    e. Coracoid process
74. Right scapula, lateral view
    a. Acromion
    b. Lateral border
    c. Glenoid cavity
    d. Coracoid process
75. Humerus, anterior view
    a. Greater tubercle
    b. Lesser tubercle
    c. Intertubercular groove
    d. Deltoid tuberosity
    e. Lateral epicondyle
    f. Capitulum
    g. Trochlea
    h. Medial epicondyle
    i. Coronoid fossa
    j. Head
76. Radius and ulna, anterior view
    a. Trochlear notch

b. Head of radius
c. Radial tuberosity
d. Styloid process of radius
e. Styloid process of ulna
f. Coronoid process
g. Olecranon process

77. Humerus, posterior view
    a. Head
    b. Anatomical neck
    c. Surgical neck
    d. Olecranon fossa
    e. Medial epicondyle
    f. Trochlea
    g. Lateral epicondyle
    h. Greater tubercle
78. Radius and ulna, posterior view
    a. Olecranon process
    b. Styloid process of ulna
    c. Styloid process of radius
    d. Radial tuberosity
    e. Neck
    f. Head of radius
79. Bones of the hand and wrist
    a. Trapezoid
    b. Trapezium
    c. Scaphoid
    d. Radius
    e. Ulna
    f. Lunate
    g. Triquetrum
    h. Capitate
    i. Pisiform
    j. Hamate

THE PELVIC GIRDLE AND LOWER EXTREMITIES

Fill in the Blanks

80. pubis
81. acetabulum
82. tibia, fibula
83. talus
84. patella

Short Answer

85. Ilium, ischium, and pubis
86. Femur, tibia, fibula
87. Calcaneus, talus, navicular, medial, intermediate, lateral cuneiform, and cuboid

Multiple Choice

88. b
89. c
90. b
91. b
92. c
93. b
94. c
95. a
96. b
97. c

Matching

98. e
99. j
100. f
101. i
102. b
103. a
104. h
105. d
106. g
107. c

Labeling

108. Left coxal bone disarticulated, lateral view
     a. Posterior superior iliac spine
     b. Posterior inferior iliac spine
     c. Acetabulum
     d. Ischial spine
     e. Ischial tuberosity
     f. Anterior inferior iliac spine
     g. Anterior superior iliac spine
     h. Iliac crest
109. Right femur, anterior and posterior views
     a. Neck
     b. Greater trochanter
     c. Lateral epicondyle
     d. Lateral condyle
     e. Medial condyle
     f. Medial epicondyle
     g. Adductor tubercle
     h. Lesser trochanter
     i. Head
     j. Intercondylar fossa
     k. Linea aspera
110. Right tibia and fibula, anterior view
     a. Lateral condyle
     b. Head of fibula
     c. Lateral malleolus
     d. Medial malleolus
     e. Crest
     f. Tibial tuberosity
     g. Medial condyle

111. Bones of the right foot, viewed from above
    a. Medial cuneiform
    b. Intermediate cuneiform
    c. Lateral cuneiform
    d. Navicular
    e. Talus
    f. Calcaneus
    g. Cuboid

ARTICULATIONS

Fill in the Blanks

112. syndesmoses
113. articulation
114. synarthroses, diarthroses
115. synchondroses
116. gliding
117. hinge, pivot
118. gomphoses

Multiple Choice

119. b
120. b
121. a
122. d
123. c
124. a
125. a
126. c

Matching

127. c
128. j
129. i
130. h
131. g
132. d
133. e
134. a
135. b
136. f

CHAPTER 10: MUSCULAR SYSTEM

INTRODUCTION TO SKELETAL MUSCLES

Fill in the Blanks

1. fibers
2. T tubules
3. sarcomere
4. sarcolemma, sodium
5. neuromuscular junction
6. fascicles, perimysium
7. aponeurosis

Matching

8. a
9. c
10. j
11. f
12. d
13. i
14. h
15. g
16. e
17. b
18. l
19. m
20. k

Multiple Choice

21. d
22. d
23. c
24. a
25. a
26. b
27. c
28. d
29. a
30. c
31. a
32. c
33. c
34. c
35. a

Labeling

36. Structures of a skeletal muscle
    a. Fascia
    b. Fascicle
    c. Myofibril
    d. Muscle fiber
    e. Endomysium
    f. Perimysium
    g. Epimysium
37. Unique features of the skeletal muscle cell
    a. Sarcoplasm
    b. Sarcolemma
    c. Myofilament
    d. T tubule
    e. Sarcoplasmic reticulum

f. Myofibril
g. A band
h. Z disk
i. Sarcomere

SKELETAL MUSCLES: ENERGY SOURCES, FUNCTION, AND STRUCTURES

Fill in the Blanks

38. myoglobin
39. fewer, finely
40. isotonic, isometric

Matching

41. h
42. f
43. e
44. c
45. d
46. b
47. g
48. a

Multiple Choice

49. d
50. c
51. b
52. a
53. d
54. b
55. d
56. a
57. b

Short Answer

58. Origin and insertion
59. Agonist, antagonist, synergists, and fixator

MUSCLES OF THE HEAD AND TRUNK

Matching

60. d
61. f
62. c
63. g
64. h
65. e
66. b
67. c
68. i
69. a
70. b
71. e
72. a
73. c
74. b
75. e
76. b
77. d
78. c
79. h
80. b
81. a
82. c
83. j
84. e
85. f
86. k
87. g
88. d
89. m
90. l
91. i

Multiple Choice

92. c
93. a
94. b
95. c
96. c
97. c
98. a
99. c
100. c
101. a
102. b
103. a
104. d

Labeling

105. Muscles of the head and neck
    a. Epicranial aponeurosis
    b. Temporalis
    c. Occipitofrontalis (occipital portion)
    d. Masseter
    e. Sternocleidomastoid
    f. Depressor anguli oris
    g. Orbicularis oris
    h. Buccinator
    i. Zygomaticus major
    j. Orbicularis oculi
    k. Corrugator supercilii

l. Occipitofrontalis (frontal portion)

106. Muscles of the back
    a. Trapezius
    b. Deltoid
    c. Infraspinatus
    d. Teres minor
    e. Teres major
    f. Latissimus dorsi
    g. External abdominal oblique
    h. Erector spinae
    i. Serratus anterior
    j. Supraspinatus
    k. Rhomboid major
    l. Rhomboid minor
    m. Levator scapulae
    n. Splenius capitis
    o. Semispinalis capitis
107. Muscles of the trunk and abdominal wall
    a. Deltoid
    b. Pectoralis major
    c. Latissimus dorsi
    d. Serratus anterior
    e. Rectus abdominis
    f. External oblique
    g. Internal oblique
    h. Transverse abdominis
108. Rotator cuff muscles
    a. Infraspinatus
    b. Teres minor
    c. Subscapularis
    d. Supraspinatus

Short Answer

109. Supraspinatus, teres minor, infraspinatus, subscapularis
110. Masseter, temporalis, pterygoid
111. Internal and external oblique, transversus abdominis, and rectus abdominis
112. Pectoralis major and latissimus dorsi

UPPER LIMB MUSCLES

Matching

113. f
114. e
115. d
116. h
117. i
118. g
119. b
120. j
121. c
122. a

Multiple Choice

123. a
124. c
125. d
126. b
127. a
128. b
129. d
130. c

Labeling

131. Muscles acting on the forearm
    a. Deltoid
    b. Biceps brachii
    c. Triceps brachii
    d. Brachialis
    e. Pronator teres
    f. Brachioradialis
    g. Teres major
    h. Coracobrachialis

LOWER LIMB MUSCLES

Matching

132. c
133. g
134. e
135. a
136. a
137. b
138. b
139. d
140. a
141. b
142. f
143. f
144. h

Multiple Choice

145. a
146. d
147. c
148. a
149. d
150. c
151. a
152. c
153. a
154. d

155. b
156. b
157. b

Labeling

158. Muscles of the anterior aspect of the thigh
    a. Gluteus medius
    b. Tensor fasciae latae
    c. Iliotibial tract
    d. Vastus lateralis
    e. Vastus medialis
    f. Rectus femoris
    g. Sartorius
    h. Adductor magnus
    i. Adductor longus
    j. Gracilis
    k. Pectineus
    l. Psoas major
    m. Iliacus
    n. Iliopsoas
159. Superficial muscles of the leg
    a. Soleus
    b. Peroneus longus
    c. Extensor digitorum longus
    d. Peroneus brevis
    e. Peroneus tertius
    f. Tibialis anterior
    g. Gastrocnemius

Short Answer

160. Adductor brevis, longus, magnus, and gracilis
161. Vastus lateralis, medialis, intermedius, and rectus femoris
162. Biceps femoris, semitendinosus, semimembranosus

CHAPTER 11: CELLS OF THE NERVOUS SYSTEM

ORGANIZATION AND CELLS OF THE NERVOUS SYSTEM, NERVE TRACTS

Fill in the Blanks

1. neurons
2. myelin, myelin
3. nodes of Ranvier
4. axon hillock
5. synapse
6. nuclei, ganglia

Matching

7. d
8. f
9. g
10. c
11. b
12. h
13. a
14. e
15. g
16. e
17. d
18. c
19. h
20. b
21. f
22. i
23. a

Multiple Choice

24. d
25. a
26. c
27. a
28. d
29. a
30. b
31. b
32. b
33. a
34. b
35. d

Labeling

36. Structure of a typical neuron
    a. Synaptic knobs
    b. Node of Ranvier
    c. Schwann cell
    d. Axon hillock
    e. Cell body or soma
    f. Dendrite

Short Answer

37. Astrocytes, microglia, ependymal, oligodendrocytes, Schwann
38. Multipolar, bipolar, unipolar
39. Afferent, efferent, interneuron
40. endoneurium, perineurium, epineurium

NERVE IMPULSES, ACTION POTENTIAL, AND SYNAPSES

Fill in the Blanks
41. nerve impulse
42. positive, negative
43. larger
44. synapse
45. anesthetics

Matching
46. c
47. e
48. a
49. g
50. f
51. b
52. i
53. d
54. h

Multiple Choice
55. a
56. b
57. a
58. a
59. d
60. b
61. b
62. c
63. a
64. d
65. a
66. c
67. b
68. c

Labeling
69. Nerve impulse conduction across the synapse
    a. Presynaptic knob
    b. Synapse
    c. Receptors
    d. Postsynaptic knob
    e. Sodium channels
    f. Acetylcholine
    g. Synaptic vesicle
    h. Depolarization
    i. Synaptic cleft

CHAPTER 12: CENTRAL NERVOUS SYSTEM

INTRODUCTION, COVERINGS OF THE BRAIN AND SPINAL CORD, CEREBROSPINAL FLUID, AND SPINAL CORD

Fill in the Blanks
1. spinal cord
2. dorsal, ventral
3. L1
4. cauda equina
5. horns, columns
6. ascending, descending
7. anterior median fissure, posterior median sulcus

Matching
8. d
9. g
10. h
11. b
12. f
13. i
14. e
15. j
16. c
17. a

Multiple Choice
18. d
19. b
20. a
21. d
22. c
23. a
24. b
25. d
26. b
27. a
28. c

Labeling
29. Coverings of the spinal cord
    a. Gray matter
    b. White matter
    c. Ventral root
    d. Dorsal root
    e. Pia mater
    f. Arachnoid mater
    g. Dura mater

h. Spinal nerve
i. Dorsal ganglion
30. Fluid spaces of the brain
a. Lateral ventricle
b. Third ventricle
c. Fourth ventricle
d. Central canal
e. Cerebral aqueduct

Short Answer
31. Dura, arachnoid, and pia mater
32. Falx cerebri, falx cerebella, tentorium cerebella
33. Epidural, subdural, subarachnoid

BRAIN AND DIENCEPHALON

Fill in the Blanks
34. basal nuclei
35. electroencephalograms (EEGs)
36. reticular activating system (RAS)
37. thalamus
38. cerebral peduncles

Matching
39. f
40. i
41. g
42. h
43. c
44. j
45. b
46. d
47. a
48. e

Multiple Choice
49. c
50. d
51. a
52. b
53. c
54. b
55. a
56. a
57. d
58. a

Labeling
59. Divisions of the brain
a. Cerebrum
b. Diencephalon
c. Thalamus
d. Pineal body
e. Hypothalamus
f. Brainstem
g. Midbrain
h. Pons
i. Medulla oblongata
j. Cerebellum

Short Answer
60. Frontal, parietal, temporal, occipital, insula
61. Medulla oblongata, pons, midbrain, cerebellum, diencephalon, and cerebrum
62. Medulla oblongata, pons, midbrain

STRUCTURES OF THE CEREBRUM AND PATHWAYS OF THE CNS

Fill in the Blanks
63. long-term memory
64. sulci, fissures
65. Broca's area
66. Primary sensory neurons or ascending tracts, somatic motor pathways

Matching
67. c
68. i
69. f
70. g
71. a
72. d
73. j
74. h
75. b
76. e

Multiple Choice
77. d
78. d
79. a
80. d
81. a
82. c
83. c
84. a
85. c
86. b
87. c

Labeling

88. Left hemisphere of cerebrum, lateral surface
    a. Central sulcus
    b. Precentral gyrus
    c. Gyri
    d. Frontal lobe
    e. Lateral fissure
    f. Temporal lobe
    g. Occipital lobe
    h. Sulci
    i. Parietal lobe
    j. Postcentral gyrus

CHAPTER 13: PERIPHERAL NERVOUS SYSTEM

SPINAL NERVES

Fill in the Blanks

1. spinal nerves, cranial nerves
2. eight, twelve, five, five, one
3. plexuses

Matching

4. e
5. a
6. h
7. b
8. c
9. i
10. d
11. f
12. j
13. g

Multiple Choice

14. b
15. c
16. b
17. a
18. b
19. c
20. d

Short Answer

21. Cervical, brachial, lumbar, sacral
22. C1-C4
23. C5-T1
24. L1-L5
25. L4, L5, S1, S4, S5

CRANIAL NERVES

Matching

26. h
27. i
28. g
29. e
30. c
31. d
32. k
33. a
34. j
35. b
36. l
37. f

Multiple Choice

38. b
39. d
40. c
41. b
42. c
43. a
44. a
45. d
46. c

Labeling

47. Cranial nerves
    a. Trochlear
    b. Optic
    c. Oculomotor
    d. Abducens
    e. Facial
    f. Vestibulocochlear
    g. Vagus
    h. Accessory
    i. Hypoglossal
    j. Glossopharyngeal
    k. Trigeminal
    l. Olfactory

DIVISIONS OF THE PERIPHERAL NERVOUS SYSTEM

Fill in the Blank

48. reflex
49. somatic
50. antagonistic
51. preganglionic
52. norepinephrine, acetylcholine

Matching
53. S
54. P
55. P
56. S
57. P
58. S
59. P
60. S
61. S
62. P

Multiple Choice
63. d
64. c
65. b
66. d
67. b
68. a
69. c
70. c
71. b
72. b
73. d

Short Answer
74. Sensory (afferent), motor (efferent)
75. Knee jerk, ankle jerk, Babinski, corneal, abdominal
76. Sympathetic, parasympathetic
77. Increased; heart rate, respiratory rate, blood pressure, blood glucose, inhibits digestion, perspiration, dilating respiratory passages
78. Decreased; heart rate, respiratory rate, blood pressure, digestion, urination, defecation

CHAPTER 14: SENSE ORGANS

RECEPTORS, SENSE OF SMELL AND TASTE

Fill in the Blanks
1. receptor
2. sensation
3. perception
4. tonic, phasic

Matching
5. d
6. f
7. e
8. d
9. b
10. c
11. e
12. f
13. a
14. a
15. e
16. a
17. c
18. d
19. f
20. b

Multiple Choice
21. d
22. a
23. b
24. c
25. b
26. c
27. a
28. b
29. c
30. b
31. d
32. a
33. c
34. b

Short Answer
35. Touch, temperature, pain, stretch, vibration, and proprioception
36. Vision, hearing, balance, taste, and smell
37. Location, type of stimulus, structure
38. Exteroceptors, interoceptors, proprioceptors
39. Mechanoreceptors, chemoreceptors, thermoreceptors, nociceptors, photoreceptors, osmoreceptors
40. Free nerve endings and encapsulated
41. Tactile, lamellar, and bulbous corpuscles
42. Sour, sweet, bitter, salty, and umami (savory)
43. Fungiform, circumvallate, foliate, filiform

SENSE OF HEARING AND BALANCE

Fill in the Blanks
44. auricle

45. spiral organ or organ of Corti
46. perilymph
47. vestibular, basilar
48. oval window, scala vestibule, vestibular, organ of Corti

Matching

49. e
50. c
51. i
52. a
53. b
54. j
55. d
56. g
57. f
58. h

Multiple Choice

59. d
60. c
61. b
62. d
63. b
64. a
65. b
66. c
67. a
68. b
69. c
70. a
71. c
72. c
73. c
74. b
75. a

Labeling

76. The ear
    a. External ear
    b. Middle ear
    c. Inner ear
    d. Auricle
    e. External acoustic meatus
    f. Tympanic membrane
77. Effect of sound waves on cochlear structures
    a. Cochlear duct
    b. Stapes
    c. Incus
    d. Malleus
    e. Tympanic membrane
    f. Oval window
    g. Round window
    h. Auditory tube
    i. Scala tympani
    j. Scala vestibule
    k. Cochlear nerve

Short Answer

78. External, middle, inner
79. Stapes, incus, malleus
80. Vestibule, cochlea, semicircular canals

VISION

Fill in the Blanks

81. canthus
82. pupil
83. retina
84. rods, cones
85. contracted, relaxed
86. opsin, retinal

Matching

87. c
88. d
89. e
90. i
91. h
92. g
93. a
94. j
95. f
96. b

Multiple Choice

97. a
98. c
99. c
100. b
101. a
102. b
103. c
104. d
105. a
106. c
107. d
108. b
109. b
110. a
111. a
112. b
113. d

114. a
115. b
116. d

Labeling

117. Lacrimal apparatus
    a. Lacrimal canals
    b. Lacrimal sac
    c. Nasolacrimal duct
    d. Puncta
    e. Lacrimal ducts
    f. Lacrimal gland
118. Horizontal section through the eyeball
    a. Pupil
    b. Lens
    c. Optic disc
    d. Optic nerve
    e. Fovea centralis
    f. Macula
    g. Posterior chamber
    h. Sclera
    i. Choroid
    j. Retina
    k. Suspensory ligament
    l. Ciliary body
    m. Iris
    n. Anterior chamber
    o. Cornea
119. Extrinsic muscles of the right eye, lateral view
    a. Trochlea
    b. Superior oblique
    c. Superior rectus
    d. Lateral rectus
    e. Inferior oblique
    f. Inferior rectus

Short Answer

120. Glands, canals, ducts, sacs, nasolacrimal ducts
121. Superior, inferior, medial, lateral rectus, superior oblique, inferior oblique
122. Fibrous, vascular, inner layer
123. Choroid, ciliary body, iris
124. Anterior cavity – aqueous humor, posterior cavity – vitreous humor
125. Cornea, aqueous humor, lens, vitreous humor

CHAPTER 15: ENDOCRINE SYSTEM

HORMONES AND PROSTAGLANDINS AND RELATED COMPOUNDS

Fill in the Blanks

1. hormones
2. steroid hormones, nonsteroid hormones
3. receptor molecules
4. target
5. plasma membrane, second messenger
6. negative feedback

Multiple Choice

7. b
8. a
9. d
10. a
11. d
12. d
13. c
14. a
15. b

Short Answer

16. Protein, glycoprotein, peptides, amino acid derivatives
17. Regulate blood pressure, metabolism, bodily functions, pain, fever, inflammation
18. Mobile-receptor model and second messenger model

THE GLANDS: PITUITARY, PINEAL, THYROID, PARATHYROID, ADRENAL, PANCREAS, GONADS, PLACENTA, THYMUS, AND OTHER ORGANS AND TISSUES THAT PRODUCE HORMONES

Fill in the Blanks

19. infundibulum
20. interstitial
21. follicles
22. progesterone

Matching

23. h
24. f
25. d
26. c
27. j

28. g
29. k
30. i
31. a
32. l
33. e
34. b
35. a
36. d
37. j
38. i
39. h
40. h
41. l
42. g
43. f
44. e
45. l
46. k
47. c
48. b
49. l
50. g
51. d
52. f
53. b
54. h
55. j
56. e
57. c
58. a
59. k
60. i
61. d
62. e
63. b
64. g
65. h
66. a
67. l
68. k
69. m
70. i
71. f
72. j
73. c

Multiple Choice

74. c
75. a
76. c
77. a
78. b
79. a
80. d
81. b
82. a
83. a
84. a
85. d
86. a
87. a
88. c

Short Answer

89. Alpha cells – glucagon, beta cells – insulin, delta cells – somatostatin, F cells – pancreatic polypeptide
90. Growth hormone (GH), prolactin (PRL), thyroid-stimulating hormone (TSH), adrenocorticotropic hormone (ACTH), follicle-stimulating hormone (FSH), and luteinizing hormone (LH)
91. ADH and oxytocin
92. Mineralocorticoids, glucocorticoids, gonadocorticoids

CHAPTER 16: BLOOD

BLOOD COMPOSITION AND RED BLOOD CELLS

Fill in the Blanks

1. plasma, formed elements
2. hematocrit
3. erythrocytes
4. oxyhemoglobin, carbaminohemoglobin
5. kidneys, bone marrow

Matching

6. c
7. g
8. b
9. d
10. h
11. f
12. a
13. i
14. e

Multiple Choice

15. c
16. b

17. d
18. a
19. c
20. d
21. b
22. c
23. d
24. a

WHITE BLOOD CELLS

Fill in the Blanks
25. white blood cells
26. T lymphocytes, B lymphocytes
27. leukopenia, leukocytosis
28. myeloid, lymphatic
29. platelets or thrombocytes

Matching
30. b
31. d
32. e
33. a
34. c
35. a
36. e
37. c
38. a
39. d

Multiple Choice
40. a
41. c
42. b
43. c
44. d
45. b
46. a
47. d
48. c
49. d
50. c
51. b
52. a
53. c
54. a

Labeling
55. Composition of whole blood
    a. Plasma
    b. Formed elements
    c. Proteins
    d. Water
    e. Platelets
    f. Leukocytes
    g. Erythrocytes
    h. Neutrophils
    i. Lymphocytes
    j. Monocytes
    k. Eosinophils
    l. Basophils
56. The formed elements of blood
    a. Thrombocytes
    b. Erythrocytes
    c. Basophil
    d. Neutrophil
    e. Eosinophil
    f. Lymphocyte
    g. Monocyte
    h. Agranulocytes
    i. Granulocytes
    j. Leukocytes

Short Answer
57. Erythrocytes, leukocytes, thrombocytes
58. Granulocytes and agranulocytes
59. Neutrophils, eosinophils, basophils, lymphocytes, monocytes
60. Neutrophils, eosinophils, basophils
61. Lymphocytes and monocytes

BLOOD TYPES, BLOOD PLASMA, AND BLOOD CLOTTING

Fill in the Blanks
62. antigens
63. 90, 10, electrolytes, nonelectrolytes
64. blood serum
65. extrinsic, intrinsic
66. vitamin K

Matching
67. e
68. h
69. d
70. c
71. f
72. b
73. i
74. g
75. a
76. j

Multiple choice

77. c
78. b
79. a
80. a
81. d
82. a
83. d
84. d
85. a
86. c
87. d
88. c
89. c
90. a
91. b
92. a
93. a
94. c

Short Answer

95. Albumins, globulins, clotting proteins
96. Viscosity, osmotic pressure, volume
97. Prothrombin, thrombin, fibrinogen, fibrin
98. Plasminogen, tissue plasminogen activator, plasmin
99. Heparin and coumarin

CHAPTER 17: ANATOMY OF THE CARDIOVASCULAR SYSTEM

HEART

Fill in the Blanks

1. mediastinum
2. pericardium
3. pericardial fluid
4. coronary
5. SA node

Matching

6. b
7. a
8. h
9. g
10. i
11. j
12. e
13. f
14. c
15. d

Multiple Choice

16. c
17. a
18. c
19. a
20. c
21. b
22. b
23. c
24. a
25. a
26. d
27. a

Labeling

28. Wall of the heart
    a. Fibrous pericardium/parietal pericardium
    b. Trabeculae carneae
    c. Endocardium
    d. Myocardium
    e. Visceral pericardium or epicardium
    f. Pericardial space/cavity
29. The heart and great vessels, anterior view
    a. Superior vena cava (SVC)
    b. Ascending aorta
    c. Pulmonary trunk
    d. Right pulmonary veins
    e. Right coronary artery
    f. Apex
    g. Anterior interventricular branches of left coronary artery and cardiac vein
    h. Circumflex artery
    i. Left pulmonary veins
    j. Left atrium
    k. Ligamentum arteriosum
    l. Arch of aorta
30. Interior of the heart
    a. Aortic (SL) valve
    b. Right atrium
    c. Left AV (mitral) valve
    d. Right AV (tricuspid) valve
    e. Chordae tendineae
    f. Interventricular septum
    g. Papillary muscle
    h. Trabeculae carneae
    i. Openings to coronary arteries
31. Coronary arteries
    a. Right coronary artery
    b. Posterior interventricular artery
    c. Right marginal artery
    d. Anterior interventricular artery

e. Left marginal artery
f. Circumflex artery
g. Left coronary artery

Short Answer
32. Fibrous pericardium and serous pericardium
33. Epicardium, myocardium, endocardium
34. Right and left atrium, right and left ventricles
35. Tricuspid, bicuspid
36. Pulmonary and aortic

BLOOD VESSELS

Fill in the Blanks
37. arteries, veins, capillaries
38. precapillary sphincters
39. collagen fibers
40. capillaries
41. muscular

Matching
42. c
43. i
44. d
45. f
46. a
47. j
48. e
49. g
50. b
51. h

Multiple Choice
52. b
53. c
54. c
55. a
56. a
57. a
58. b
59. c

Labeling
60. Structure of blood vessels
   a. Valve
   b. Tunica intima
   c. Endothelium
   d. Basement membrane
   e. Tunica media
   f. Tunica externa
   g. Vein
   h. Capillary
   i. Artery

Short Answer
61. Arteries, capillaries, veins
62. Elastic arteries, muscular arteries, arterioles, metarterioles
63. Endothelial cells, collagen fibers, elastic fibers, smooth muscle cells
64. Tunica intima, tunica media, tunica externa

MAJOR BLOOD VESSELS – ARTERIES

Fill in the Blanks
65. systemic circulation
66. pulmonary circulation
67. vascular anastomosis
68. aorta

Matching
69. e
70. c
71. g
72. i
73. b
74. a
75. d
76. f
77. j
78. h

Multiple Choice
79. a
80. b
81. b
82. d
83. d
84. b
85. a
86. b
87. a

Labeling
88. Major arteries of the head and neck
   a. Internal carotid
   b. External carotid
   c. Vertebral
   d. Common carotid
   e. Subclavian
   f. Brachiocephalic

89. Divisions and primary branches of the aorta
    a. Right common carotid
    b. Right subclavian
    c. Brachiocephalic
    d. Ascending aorta
    e. Left coronary
    f. Right coronary
    g. Abdominal aorta
    h. Median sacral
    i. Left femoral
    j. External iliac
    k. Internal iliac
    l. Common iliac
    m. Inferior mesenteric
    n. Gonadal
    o. Superior mesenteric
    p. Renal
    q. Suprarenal
    r. Splenic
    s. Celiac trunk
    t. Aortic arch
    u. Left subclavian
    v. Left common carotid
90. Principal arteries of the body
    a. Right common carotid
    b. Subclavian
    c. Brachiocephalic
    d. Axillary
    e. Brachial
    f. Popliteal
    g. Anterior tibial
    h. Posterior tibial
    i. Dorsal pedis
    j. Femoral
    k. Digital
    l. Superficial palmar arch
    m. Deep palmar arch
    n. Ulnar
    o. Radial
    p. External carotid
    q. Internal carotid

Short Answer

91. Brachial, radial, ulnar, palmar arches, digital
92. Common carotid, internal carotid, external carotid, vertebral
93. Celiac, superior mesenteric, inferior mesenteric, suprarenal, renal, ovarian, testicular
94. Femoral, popliteal, anterior tibial, posterior tibial, plantar arch, digital
95. Ascending, aortic arch, descending, thoracic, abdominal

VEINS AND FETAL CIRCULATION

Fill in the Blanks

96. umbilical
97. ductus venosus
98. superior vena cava, inferior vena cava
99. azygos

Matching

100. d
101. e
102. a
103. j
104. b
105. c
106. g
107. i
108. e
109. h

Multiple Choice

110. b
111. b
112. c
113. b
114. a
115. d
116. d
117. a
118. d
119. a
120. c
121. a
122. b

Labeling

123. Inferior vena cava and its abdominopelvic tributaries
    a. Right brachiocephalic
    b. Superior vena cava
    c. Azygos
    d. Phrenic
    e. Hepatic
    f. Inferior vena cava
    g. Right renal
    h. Right ovarian or testicular (gonadal)
    i. Right common iliac

j. Right external iliac
k. Right internal iliac

124. Principal veins of the body
 a. Right brachiocephalic
 b. Right subclavian
 c. Superior vena cava (SVC)
 d. Inferior vena cava (IVC)
 e. Hepatic
 f. Hepatic portal
 g. Superior mesenteric
 h. Common iliac
 i. External iliac
 j. Femoral
 k. Great saphenous
 l. Small saphenous
 m. Fibular
 n. Anterior tibial
 o. Posterior tibial
 p. Venous dorsal arch
 q. Digital
 r. Popliteal
 s. Radial
 t. Ulnar
 u. Basilic
 v. Cephalic
 w. Axillary
 x. Left subclavian
 y. Left brachiocephalic
 z. Internal jugular
 aa. External jugular
125. Plan of fetal circulation
 a. Ductus arteriosus
 b. Pulmonary trunk
 c. Superior vena cava
 d. Inferior vena cava
 e. Ductus venosus
 f. Hepatic portal vein
 g. Placenta
 h. Umbilical cord
 i. Umbilical arteries
 j. Umbilical vein
 k. Internal iliac arteries

Short Answer

126. Cephalic, brachial, radial, ulnar, basilic, median cubital, palmar arches, digital
127. Cystic, gastric, splenic, inferior mesenteric, pancreatic, superior mesenteric, gastroepiploic
128. Lingual, facial, superior thyroid, sigmoid sinus, cavernous sinus, ophthalmic, transverse sinus
129. Occipital sinus, straight sinus, superior sagittal sinus

CHAPTER 18: PHYSIOLOGY OF THE CARDIOVASCULAR SYSTEM

HEMODYNAMICS AND THE HEART AS A PUMP

Fill in the Blanks

1. hemodynamics
2. subendocardial branches
3. electrocardiograph
4. cardiac cycle
5. ventricular diastole, atrioventricular
6. heart murmur

Matching

7. h
8. f
9. d
10. g
11. e
12. i
13. j
14. c
15. a
16. b

Multiple Choice

17. d
18. b
19. b
20. d
21. a
22. d
23. b
24. a
25. b
26. a
27. c

Labeling

28. Conduction system of the heart
 a. AV node
 b. SA node
 c. Internodal bundles
 d. AV bundle
 e. Right and left AV bundle branches
 f. Subendocardial fibers or Purkinje fibers

29. Electrocardiogram deflection waves
    a. R
    b. P wave
    c. Q
    d. S
    e. ST interval
    f. T wave

PRIMARY PRINCIPLES OF CIRCULATION

Fill in the Blanks
30. cardiac output, peripheral resistance
31. stroke volume (SV), peripheral resistance
32. parasympathetic
33. viscosity
34. vasoconstriction, vasodilation
35. decrease, vasoconstriction

Matching
36. h
37. a
38. b
39. j
40. d
41. f
42. i
43. g
44. e
45. c

Multiple Choice
46. b
47. a
48. c
49. b
50. b
51. d
52. b
53. c
54. a
55. b
56. a
57. b
58. d
59. c
60. c

Short Answer
61. Volume, CO, resistance
62. Aortic and carotid
63. Emotions, exercise, hormones, blood temperature, pain, exteroceptors
64. Friction and viscosity
65. Gravity, respirations, skeletal muscle contraction
66. Radial, temporal, carotid, brachial, posterior tibial, dorsalis pedis, subclavian, femoral, axillary

CHAPTER 19: LYMPHATIC AND IMMUNE SYSTEMS

LYMPHATIC SYSTEM

Fill in the Blanks
1. lymphatic system, interstitial
2. endothelial
3. chyle
4. sinuses
5. lymph nodes
6. white pulp, red pulp

Matching
7. j
8. i
9. h
10. a
11. b
12. c
13. d
14. f
15. e
16. g

Multiple Choice
17. a
18. b
19. a
20. c
21. b
22. c
23. c
24. c
25. c
26. c
27. b
28. d
29. a
30. c
31. c
32. c

Labeling

33. Principal organs of the lymphatic system
    a. Tonsils
    b. Cervical lymph node
    c. Right lymphatic duct
    d. Inguinal lymph node
    e. Spleen
    f. Thoracic duct
    g. Axillary lymph node
    h. Thymus gland
34. Circulation plan of lymphatic fluid
    a. Lymphatic capillaries
    b. Lymph node
    c. Lymphatic vessels
    d. Efferent lymphatic vessel
    e. Afferent lymphatic vessel
    f. Interstitial fluid
    g. Lymphatic fluid
35. Location of the tonsils
    a. Pharyngeal tonsil
    b. Palatine tonsil
    c. Lingual tonsil

Short Answer

36. Maintains fluid balance, serves as part of the immune system, helps absorption of lipids from intestines
37. Breathing, skeletal muscle contraction, valves
38. Defense, hematopoiesis, RBC and platelet destruction, reservoir for blood
39. Palatine, pharyngeal, lingual
40. Lymph nodes, tonsils, thymus gland, spleen, bone marrow

IMMUNE SYSTEM

Fill in the Blanks

41. antigens
42. self-tolerance
43. phagocytosis
44. pyrogen

Matching

45. c
46. j
47. h
48. e
49. d
50. f
51. a
52. g
53. i
54. b

Multiple Choice

55. b
56. c
57. c
58. a
59. a
60. a
61. b
62. a
63. c
64. b
65. d

Short Answer

66. Epithelial barrier cells, phagocytic cells, natural killer cells
67. Interleukins, leukotrienes, interferons
68. Histamine, kinins, prostaglandins, leukotrienes, interleukins
69. Mechanical and chemical barriers, inflammation, fever, phagocytosis, NK cells, interferon, complement proteins

ADAPTIVE IMMUNITY

Fill in the Blanks

70. B lymphocytes, T lymphocytes
71. antibodies, antigens
72. immunoglobulins
73. antigen-antibody complex
74. effector, memory

Matching

75. i
76. c
77. b
78. a
79. d
80. g
81. f
82. h
83. e

Multiple Choice

84. d
85. a
86. c
87. b
88. d

89. c
90. d
91. c
92. a
93. c
94. a
95. b
96. b
97. a
98. a
99. a
100. b
101. c

Short Answer

102. G, A, M, E, D
103. inactivates toxins, agglutination, activates complement proteins, targets cells for phagocytosis, can cause lyses of the cell
104. Natural passive, natural active, artificial passive, artificial active

CHAPTER 20: RESPIRATORY SYSTEM

UPPER RESPIRATORY TRACT

Fill in the Blanks

1. meatuses
2. vestibule
3. paranasal sinuses
4. larynx

Matching

5. d
6. f
7. g
8. i
9. c
10. a
11. b
12. j
13. e
14. h

Multiple Choice

15. c
16. a
17. d
18. a
19. b
20. a
21. c
22. b
23. d
24. a
25. c
26. a
27. a
28. c

Labeling

29. Upper respiratory tract
    a. Frontal sinus
    b. Superior nasal concha
    c. Middle nasal concha
    d. Vestibule
    e. Inferior concha
    f. Anterior naris or nostril
    g. Hyoid bone
    h. Larynx
    i. Trachea
    j. Esophagus
    k. Laryngopharynx
    l. Epiglottis
    m. Oropharynx
    n. Nasopharynx
    o. Opening of auditory tube
    p. Sphenoid sinus
30. Larynx
    a. Epiglottis
    b. Vestibular fold
    c. Hyoid bone
    d. Ventricle
    e. Vocal fold
    f. Thyroid cartilage
    g. Cricoid cartilage
    h. Tracheal rings/cartilages of trachea
    i. Corniculate cartilage
    j. Cuneiform cartilage
    k. Vestibule
31. Laryngeal cartilages
    a. Hyoid bone
    b. Thyroid cartilage
    c. Cricoid cartilage
    d. Trachea
    e. Arytenoid cartilage
    f. Corniculate cartilage
    g. Epiglottis
32. Vocal folds
    a. Epiglottis
    b. Glottis
    c. Vestibular fold
    d. Cuneiform cartilage

e. Corniculate cartilage
f. Arytenoid cartilage
g. Vocal folds

Short Answer

33. Gas distributor/exchanger; filter, warm, and humidify air; communication; olfaction; regulating homeostasis
34. Nose, nasopharynx, oropharynx, laryngopharynx, larynx
35. Trachea, bronchial tree, lungs
36. Frontal, maxillary, ethmoid, sphenoid
37. Nasopharynx, oropharynx, laryngopharynx
38. Thyroid, cricoid, epiglottis, arytenoid, corniculate, cuneiform

LOWER RESPIRATORY TRACT

Fill in the Blanks

39. trachea
40. surface tension
41. pseudostratified ciliated columnar
42. right, left

Matching

43. d
44. c
45. f
46. h
47. b
48. a
49. i
50. g
51. e
52. j

Multiple Choice

53. a
54. c
55. b
56. b
57. a

Labeling

58. Respiratory system
    a. Upper respiratory tract
    b. Lower respiratory tract
    c. Bronchioles
    d. Alveoli
    e. Left and right primary bronchi
    f. Trachea
    g. Larynx
    h. Laryngopharynx
    i. Oropharynx
    j. Nasopharynx
    k. Pharynx
59. Lungs and pleura, transverse section
    a. Primary bronchus
    b. Visceral pleura
    c. Parietal pleura
    d. Intrapleural space

Short Answer

60. Primary, secondary, tertiary, bronchioles, terminal bronchioles, respiratory bronchioles, alveolar sacs.

RESPIRATORY PHYSIOLOGY

Fill in the Blanks

61. expands, gradient, higher, into, principle of ventilation
62. alveolar air, lung capillaries
63. dissolved gas, bicarbonate

Matching

64. e
65. c
66. a
67. b
68. j
69. g
70. d
71. i
72. f
73. h

Multiple Choice

74. a
75. d
76. c
77. a
78. b
79. d
80. d
81. b
82. b
83. a

Labeling

84. Pulmonary ventilation volumes and capacities
    a. Total lung capacity (TLC)

b. Residual volume (RV)
c. Expiratory reserve volume (ERV)
d. Tidal volume (TV)
e. Inspiratory reserve volume (IRV)
f. Vital capacity (VC)

CHAPTER 21: DIGESTIVE SYSTEM

MOUTH TO STOMACH

Fill in the Blanks

1. alimentary, gastrointestinal (GI) tract
2. mastication, deglutition
3. bolus

Matching

4. h
5. f
6. e
7. g
8. i
9. j
10. d
11. a
12. b
13. c

Multiple Choice

14. a
15. a
16. a
17. a
18. b
19. d
20. c
21. b
22. a
23. c
24. d
25. a

Labeling

26. Location of digestive organs
    a. Parotid gland
    b. Submandibular gland
    c. Pharynx
    d. Esophagus
    e. Diaphragm
    f. Colon or large intestine
    g. Rectum
    h. Anal canal
    i. Small intestine
    j. Spleen
    k. Stomach
    l. Liver
    m. Trachea
    n. Sublingual gland
27. The oral cavity
    a. Hard palate
    b. Soft palate
    c. Tongue
    d. Palatine tonsil
    e. Uvula
28. Typical tooth
    a. Crown
    b. Neck
    c. Root
    d. Bone (alveolar process)
    e. Cementum
    f. Periodontal ligament
    g. Root canal
    h. Gingiva
    i. Pulp cavity
    j. Dentin
    k. Enamel
29. Stomach
    a. Esophagus
    b. Lower esophageal sphincter (LES)
    c. Cardia
    d. Lesser curvature
    e. Pylorus
    f. Pyloric sphincter
    g. Duodenum
    h. Rugae
    i. Greater curvature
    j. Mucosa
    k. Submucosa
    l. Oblique muscle
    m. Circular muscle
    n. Longitudinal muscle
    o. Body of stomach
    p. Fundus

Short Answer

30. Mucosa, submucosa, muscularis, serosa
31. Lips, cheeks, hard and soft palate, tongue
32. Sublingual, submandibular, parotid
33. Crown, neck, root
34. Pylorus, body, fundus
35. Food reservoir, secretion of gastric juice, make intrinsic factor, produce hormones, protection

SMALL INTESTINE TO PANCREAS

Fill in the Blanks

36. duodenum
37. mesentery
38. pancreatic islets

Matching

39. c
40. i
41. g
42. d
43. e
44. a
45. b
46. h
47. f
48. j

Multiple Choice

49. d
50. d
51. d
52. a
53. b
54. a
55. d
56. a
57. c

Labeling

58. Wall of the small intestine
    a. Microvilli
    b. Epithelial cells
    c. Villi
    d. Plica
    e. Mucosa
    f. Submucosa
    g. Muscularis
    h. Serosa
    i. Mesentery
59. Divisions of the large intestine
    a. Hepatic flexure
    b. Ascending colon
    c. Ileocecal valve
    d. Cecum
    e. Vermiform appendix
    f. Ileum
    g. Mesentery
    h. Rectum
    i. External anal sphincter
    j. Anus
    k. Sigmoid colon
    l. Haustra
    m. Descending colon
    n. Taeniae coli
    o. Splenic flexure
    p. Transverse colon
60. The rectum and anus
    a. Rectum
    b. Anal column
    c. Anal canal
    d. Anus
    e. Rectal valves
61. Structure of the liver
    a. Inferior vena cava
    b. Right lobe
    c. Gallbladder
    d. Round ligament
    e. Falciform ligament
    f. Left lobe
    g. Caudate lobe
    h. Quadrate lobe
62. Bile ducts
    a. Gallbladder
    b. Cystic duct
    c. Liver
    d. Major duodenal papilla
    e. Duodenum
    f. Pancreas
    g. Pancreatic duct
    h. Common bile duct
    i. Common hepatic duct
    j. Right and left hepatic ducts

Short Answer

63. Ileum, jejunum, duodenum
64. Cecum, ascending colon, hepatic flexure, transverse colon, splenic flexure, descending colon, sigmoid, rectum, anal canal
65. Right, left, caudate, quadrate
66. Detoxify, metabolism, store vitamins and minerals, produce plasma proteins, secrete bile
67. Acinar, beta, alpha

DIGESTIVE FUNCTION

Fill in the Blanks

68. mechanical digestion
69. peristalsis
70. segmentation

71. chemical digestion
72. amylases, proteases, lipases

Matching
73. c
74. g
75. a
76. c
77. b
78. i
79. h
80. f
81. j
82. e

Multiple Choice
83. b
84. a
85. d
86. c
87. a
88. a
89. a
90. b
91. a
92. c
93. d
94. a
95. d

Short Answer
96. Sucrase, lactase, maltase
97. Pepsin, trypsin, chymotrypsin, peptidases
98. Cephalic, gastric, and intestinal
99. Oral, pharyngeal, esophageal
100. Protease, amylase, lipase

CHAPTER 22: NUTRITION AND METABOLISM

Fill in the Blanks
1. nutrition, metabolism
2. glucose phosphorylation, phosphorylation
3. citric acid cycle
4. lipids
5. increase, decrease

Matching
6. c
7. f
8. j
9. a
10. b
11. h
12. g
13. d
14. e
15. i

Multiple Choice
16. c
17. a
18. c
19. a
20. c
21. a
22. b
23. c
24. a
25. b
26. c
27. a
28. a
29. c
30. c
31. d
32. b
33. d
34. a

Short Answer
35. Macronutrients and micronutrients
36. Monosaccharides, disaccharides, polysaccharides
37. Chylomicrons, lipoproteins, free fatty acids
38. Very-low-density lipoproteins (VLDL), low-density lipoproteins (LDL), high-density lipoproteins (HDL)
39. Triglycerides, cholesterol, phospholipids, prostaglandins
40. Insulin, adrenocorticotropic hormone (ACTH), growth hormone (GH), glucocorticoids
41. Fever, drugs, physiological state, sex, age, pregnancy, lactation
42. Carbohydrates, fats, proteins, water, minerals
43. Vitamins and some minerals (trace elements)

CHAPTER 23: URINARY SYSTEM AND FLUID BALANCE

ANATOMY OF THE URINARY SYSTEM

Fill in the Blanks
1. retroperitoneal
2. transitional epithelium
3. trigone
4. fenestrations
5. cortical, juxtamedullary

Matching
6. b
7. e
8. c
9. f
10. g
11. a
12. d
13. h
14. k
15. i
16. j
17. d
18. g
19. f
20. e
21. j
22. b
23. k
24. a
25. c
26. h
27. i

Multiple Choice
28. c
29. a
30. d
31. b
32. b
33. a
34. b
35. c
36. b
37. c
38. d

Labeling
39. Location of the urinary system organs
    a. Adrenal gland
    b. Liver
    c. Kidneys
    d. Ureters
    e. Urinary bladder
    f. Urethra
40. Internal structure of the kidney
    a. Renal column
    b. Renal sinus
    c. Hilum
    d. Renal papilla
    e. Ureter
    f. Renal medulla
    g. Medullary pyramid
    h. Renal pelvis
    i. Fat
    j. Major calyces
    k. Minor calyces
    l. Renal cortex
41. Circulation of blood through the kidney
    a. Interlobular arteries and veins
    b. Interlobar arteries and veins
    c. Lobar arteries and veins
    d. Segmental arteries and veins
    e. Renal artery and vein
    f. Arcuate arteries and veins
42. Nephron
    a. Distal convoluted tubule (DCT)
    b. Collecting duct (CD)
    c. Papilla of renal pyramid
    d. Thin ascending limb of Henle (tALH)
    e. Henle loop
    f. Thick ascending limb of Henle (TAL)
    g. Descending limb of Henle (DLH)
    h. Proximal convoluted tubule (PCT)
    i. Renal tubule
    j. Renal corpuscle
    k. Nephron
43. Structure of renal corpuscle
    a. Efferent arteriole
    b. Distal convoluted tubule (DCT)
    c. JG apparatus
    d. Afferent arteriole
    e. Renal capsule
    f. Glomerulus
    g. Bowman capsule
    h. Proximal convoluted tubule (PCT)
    i. Podocytes

Short Answer
44. Glomerulus, Bowman capsule
45. Proximal convoluted tubule (PCT), Henle loop, distal convoluted tubule (DCT)

46. Cortical and juxtamedullary
47. Renal corpuscle, proximal convoluted tubule (PCT), Henle loop, distal convoluted tubule (DCT)

PHYSIOLOGY OF THE URINARY SYSTEM

Matching
48. f
49. h
50. a
51. g
52. e
53. d
54. c
55. b
56. i

Multiple Choice
57. d
58. a
59. d
60. a
61. c
62. a
63. d
64. b
65. c

Short Answer
66. Filtration, tubular reabsorption, tubular secretion
67. Hormone production, electrolyte balance, acid/base balance, excretion of waste, fluid balance, synthesize vitamin D
68. Rennin, antidiuretic hormone (ADH), atrial natriuretic hormone (ANH), aldosterone
69. Nitrogen wastes, electrolytes, toxins, pigments, hormones

FLUID AND ELECTROLYTE BALANCE

Fill in the Blanks
70. lower, higher
71. extracellular, intracellular
72. osmoreceptors

Matching
73. d
74. c
75. h
76. b
77. f
78. g
79. e
80. a
81. j
82. i

Multiple Choice
83. c
84. b
85. a
86. b
87. b
88. a

Short Answer
89. Age, gender, body fat
90. ICF, ECF, plasma or intravascular, and interstitial
91. Sodium, calcium, potassium, magnesium
92. Kidney, lungs, skin, intestines

CHAPTER 24: MALE REPRODUCTIVE SYSTEM

Fill in the Blanks
1. urogenital, anal
2. spermatogenesis
3. seminiferous, sperm, interstitial, testosterone
4. seminal vesicles
5. cremaster muscle
6. penis

Matching
7. c
8. h
9. f
10. i
11. k
12. d
13. b
14. a
15. l
16. g
17. e
18. j

Multiple Choice

19. b
20. c
21. a
22. d
23. a
24. c
25. a
26. b
27. a
28. a
29. d
30. b
31. b
32. c

Labeling

33. Male reproductive organs
    a. Seminal vesicle
    b. Ejaculatory duct
    c. Prostate gland
    d. Rectum
    e. Bulbourethral gland
    f. Epididymis
    g. Testis
    h. Scrotum
    i. Prepuce
    j. Glans penis
    k. Urethra
    l. Corpora cavernosa
    m. Corpus spongiosum
    n. Vas deferens
    o. Urinary bladder
    p. Ureter
34. The male reproductive system
    a. Ureter
    b. Urinary bladder
    c. Prostate gland
    d. Bulbourethral gland
    e. Vas deferens
    f. Epididymis
    g. Testis
    h. Glans penis
    i. Scrotum
    j. Tunica vaginalis
    k. Cremaster muscle
    l. Corpus spongiosum
    m. Corpora/corpus cavernosa
    n. Spermatic cord
    o. Inguinal canal
    p. Ejaculatory duct
    q. Seminal vesicle
35. Testis and epididymis
    a. Vas deferens
    b. Septum
    c. Lobule
    d. Tunica albuginea
    e. Rete testis
    f. Seminiferous tubules
    g. Efferent ductules
    h. Epididymis

Short Answer

36. Nerves, blood vessels, lymphatics
37. Spermatogenesis and secretion of testosterone
38. Head, acrosome, midpiece, tail
39. Seminal vesicles, prostate, bulbourethral glands
40. Corpora cavernosa, corpus spongiosum, glans penis, prepuce, urethra

CHAPTER 25: FEMALE REPRODUCTIVE SYSTEM

Fill in the Blanks

1. granulosa, oocyte, antrum
2. primary, mature or graafian
3. ovulation
4. uterine tubes
5. corpus luteum
6. prolactin, oxytocin

Matching

7. f
8. b
9. e
10. g
11. a
12. h
13. d
14. i
15. c
16. i
17. a
18. d
19. e
20. j
21. g
22. c
23. b
24. h
25. f

Multiple Choice

26. c
27. a
28. c
29. b
30. a
31. a
32. b
33. d
34. c
35. a
36. d
37. c
38. a
39. a
40. d
41. b
42. b

Labeling

43. Female reproductive organs
    a. Uterine tube
    b. Ovary
    c. Ureter
    d. Uterosacral ligament
    e. Cervix
    f. Fornix
    g. Vagina
    h. Labium majus
    i. Labium minus
    j. Clitoris
    k. Urethra
    l. Round ligament
    m. Fundus
    n. Body of uterus
    o. Ovarian ligament
    p. Suspensory ligament
44. Internal female reproductive organs
    a. Fundus
    b. Body of uterus
    c. Endometrium
    d. Myometrium
    e. Perimetrium
    f. Cervix
    g. Fornix
    h. External os
    i. Vagina
    j. Broad ligament
    k. Ovary
    l. Fimbriae
    m. Infundibulum
    n. Ampulla
    o. Ovarian ligament
45. Stages of ovarian follicle development
    a. Corpus luteum
    b. Ovulation
    c. Mature follicle or graafian follicle
    d. Antrum
    e. Granulosa cells
    f. Secondary follicle
    g. Primary follicle
    h. Corpus albicans
46. Vulva
    a. Prepuce
    b. Clitoris
    c. Labium minus
    d. External urinary meatus
    e. Vestibule
    f. Hymen
    g. Orifice of vagina
    h. Labium majus
    i. Mons pubis
47. The female breast
    a. Pectoralis major muscle
    b. Alveolus
    c. Lactiferous duct
    d. Lactiferous sinus
    e. Nipple pores
    f. Adipose tissue
    g. Suspensory ligaments
    h. Nipple
    i. Areola

Short Answer

48. Follicular phase, luteal phase
49. Broad, uterosacral, posterior, anterior, round
50. Fundus, body, cervix
51. Perimetrium, myometrium, endometrium
52. Menses, proliferative/postmenstrual, ovulation, secretory/premenstrual
53. Thickening of endometrium, growth of glands in endometrium, increase water content of endometrium, increase myometrial contractions
54. Preparation of the endometrium for implantation, increase water content of endometrium, decrease myometrial contractions

CHAPTER 26: GROWTH AND DEVELOPMENT

Fill in the Blanks

1. spermatogenesis, oogenesis
2. meiosis
3. tetrads, crossing over
4. granulosa
5. blastocyst
6. stem cells
7. high, human chorionic gonadotropin (hCG), oxytocin, amplifies or increases

Matching

8. j
9. g
10. h
11. b
12. c
13. e
14. a
15. d
16. f
17. i
18. f
19. d
20. b
21. i
22. h
23. g
24. j
25. e
26. a
27. c

Multiple Choice

28. c
29. a
30. a
31. c
32. b
33. b
34. d
35. b
36. b
37. d
38. b
39. a
40. d
41. a
42. b
43. d
44. b
45. c
46. a
47. d
48. c
49. c
50. d
51. a

Short Answer

52. Endoderm, mesoderm, ectoderm
53. Epidermis, enamel of teeth, cornea of eye, brain and spinal cord
54. Glands, lining of respiratory and GI tract, lining of pancreatic, hepatic, and urinary tract
55. Dermis, muscles, bones, organs, circulation
56. 1 – contractions and dilation, 2 – baby leaves mother's body, 3 – expulsion of placenta
57. Infancy, childhood, adolescence/adulthood, older adulthood

CHAPTER 27: HUMAN GENETICS AND HEREDITY

Fill in the Blanks

1. genetics
2. chromatin, chromosome
3. genome
4. dominant, recessive
5. monogenic
6. codominance
7. genetic mutation
8. p-arm, q-arm
9. X
10. amniocentesis, chorionic villus sampling

Matching

11. f
12. j
13. a
14. c
15. d
16. g
17. e
18. i
19. b
20. h

Multiple Choice

21. a
22. d
23. c
24. b
25. b
26. a
27. b
28. a
29. b
30. c
31. a
32. b
33. d
34. b
35. a
36. b
37. a